Product Decoration Technologies

Understanding the primary methods
for decorating a product

Other Labels & Labeling books:

ENCYCLOPEDIA OF LABEL TECHNOLOGY
Michael Fairley

THE HISTORY OF LABELS
Michael Fairley and Tony White

DIGITAL LABEL AND PACKAGE PRINTING
Michael Fairley

ENVIRONMENTAL PERFORMANCE AND SUSTAINABLE LABELING
Michael Fairley and Danielle Jerschefske

CONVENTIONAL LABEL PRINTING PROCESSES
John Morton and Robert Shimmin

LABEL DESIGN AND ORIGINATION
John Morton and Robert Shimmin

LABEL DISPENSING AND APPLICATION TECHNOLOGY
Michael Fairley

CODES AND CODING TECHNOLOGY
Michael Fairley

LABEL EMBELLISHMENTS AND SPECIAL APPLICATIONS
John Morton and Robert Shimmin

BRAND PROTECTION, SECURITY LABELING AND PACKAGING
Jeremy Plimmer

DIE-CUTTING AND TOOLING
Michael Fairley

MANAGEMENT INFORMATION SYSTEMS AND WORKFLOW AUTOMATION
Michael Fairley

SHRINK SLEEVE TECHNOLOGY
Michael Fairley and Séamus Lafferty

LABEL MARKETS AND APPLICATIONS
John Penhallow

For the latest list please visit: **www.labelsandlabeling.com**

Product Decoration Technologies

Understanding the primary methods for decorating a product

John Morton and Robert Shimmin
4impression

Product Decoration Technologies
Understanding the primary methods for decorating a product

First edition published 2018 by:
Tarsus Exhibitions & Publishing Ltd

Printed by CreateSpace, an Amazon.com company.

ISBN 978-1-910507-15-5

Contents

Foreword

A trip to any supermarket will reveal shelves full of colorful packaging all competing for the attention of the shopper.

The myriad of bottles, jars, sachets, packs and cartons on display are typically decorated with attractive graphics and content, informing the consumer about the contents, nature and/or purpose of the items inside.

A closer inspection will often reveal that in some instances the pack itself is printed with visual information, whereas in other cases a separate label or sleeve is the main carrier of information. In many instances, it can often be unclear as to how a product has been decorated.

The aim of this book therefore, is to look below the surface and explore the wide variety of product decoration technologies in use, highlight their unique characteristics and help in their identification. As the chapters unfold the reader will gain a comprehensive insight into the whole decorative process as it relates to consumer packaging… from material selection, print and finally to the application onto/or integration into the pack itself.

The pros and cons of each decoration system will be identified but more importantly the cost implications underpinning the selection of a particular decoration system will be examined.

John Morton and Robert Shimmin
4impression

About the Label Academy

This book is part of the recommended study material for the Label Academy, a global training and certification program for the label industry. The Label Academy was created by the team behind Labels & Labeling magazine and the Labelexpo series of events.

The Academy consists of a series of self-study modules, combining free access to relevant articles and videos with paid text books (both printed and electronic). Once a student has completed a module, there is an opportunity to take an online test and earn a certificate.

It is expected that a Label Academy qualification will become a standard in the industry – for printers/converters, suppliers, brand owners and designers – and assist in providing a benchmark. In addition to its own training, the Label Academy will aim to become a resource provider to the many existing educational programs in the industry. Accredited training courses will be promoted through the Label Academy website and books will be provided at discounted rates.

The Label Academy concept was pioneered by industry expert Mike Fairley. This was in response to a reduction in the number of dedicated printing colleges and the need to standardize training across the world. The label industry also has its own specific training needs – it has some of the widest range of materials, printing processes and finishing solutions of any printing sector.

We are also working with other training experts and authors to ensure that the Label Academy provides up-to-date and relevant training material for the industry.

The Label Academy is supported by the key trade associations, including FINAT, TLMI and the LMAI.

www.label-academy.com

Label Academy sponsors

Thank you to our founding sponsors, without whom this ambitious project would not have been possible:

Cerm

Cerm designs business automation software solutions to meet the specific demands of flexo and digital narrow web printers. Using the latest technology, our team's focus is on innovation and continuous improvement.

Our automation solutions support each step in the printer's integrated workflow – from estimating to production, shipment and data collection – and provide the feature and functionality printers need to gain efficiency and improve profitability.

Cerm inspires collaboration and helps printers remain competitive in the market and deliver the best products possible. We are proud to sponsor the Label Academy and contribute to the future of the narrow web printing industry.

www.cerm.net

Flint Group Narrow Web

Flint Group Narrow Web has the products, the solutions, and the technical experts to handle any print situation. Providing solutions for food packaging, sustainability, increased bottom line, efficiency, and uptime – delivering the basics needed to run a successful operation, and the expertise to go above and beyond to another level of success.

Our experts provide solutions to your printing problems with the innovative products and services that have made us an industry leader around the world. Wherever you are, we are – available to help you reach your business goals today and into the future.

Continuous improvement is paramount to Flint Group; we are proud to sponsor the Label Academy and the benefits it will bring to the future of our industry.

www.flintgrp.com

Gallus Group

The Gallus Group with its production sites in Switzerland and Germany is a leader in the development, production and sale of narrow-web, reel-fed presses designed for label manufacturers. The machine portfolio is augmented by a broad range of screen printing plates (Gallus Screeny), globally decentralized service operations, and a broad offering of printing accessories and replacement parts. The comprehensive portfolio also includes consulting services provided by label experts in all relevant printing and process engineering tasks. The Gallus Group is a member of the Heidelberg Group and employs around 430 people, of whom 253 are based in Switzerland. The group headquarters is in St.Gallen, Switzerland.

www.gallus-group.com

MPS Systems B.V.

Producing high-quality label printing depends on several factors; one of them is the operator of the press.

As a press machine builder since 1996, MPS Systems B.V. knows how important training and education on subjects like pre-press, label printing and finishing is. For label printers, it is critical that their operators keep up with pre-press and press developments in addition to label trends. Therefore, MPS sponsors the Label Academy, to advance operator's passion for printing, share expertise and help multiply benefits.

The MPS slogans of 'Printers First' and 'Technology with Respect' have always underlined the core philosophy of MPS from press design to operator satisfaction. We develop our presses with a strong focus on user-friendliness and respect for the press operator: Printers First.

www.mps4u.com

HP Indigo

HP Indigo is a global leader in digital printing, with a broad portfolio of digital presses and workflow solutions. Indigo's proprietary Liquid Electrophotography (LEP) technology delivers exceptional print quality for the widest variety of applications including labels, flexible packaging, shrink sleeves and folding cartons. HP Indigo's digital presses match gravure print quality satisfying the most demanding brands.

A division of HP Inc.'s Graphics Solutions Business, Indigo serves customers in more than 122 countries, including many of the top label and packaging converters worldwide.

www.hp.com/go/labelsandpackaging

UPM Raflatac

In a little more than three decades, UPM Raflatac has become one of the world's leading manufacturers of pressure sensitive label materials, developing and leveraging the latest innovations in adhesive technology. Our film and paper label stocks are used for product and information labeling across a wide range of end-uses – from pharmaceuticals and security to food and beverage applications.

We are an engineering driven company with industry-leading products known for their consistent high quality and top performance. We are also known for the high performing supply chain and undisputed leadership in the area of sustainability. UPM Raflatac's dedication to innovation, sustainability and top quality is matched only by our commitment to service excellence. We call it the Raflatouch.

www.upmraflatac.com

About the authors

4impression Training

4impression are specialist providers of training across a wide range of print and packaging related subjects. Staffed by industry trained tutors and supported by a network of print and packaging suppliers, the company delivers face to face to courses providing understanding of print processes, embellishments, materials, origination and finishing. Recently 4impression wrote the FINAT Educational Handbook which covers all aspects of self-adhesive label manufacture. They have also produced a comprehensive range of learning resources for the FINAT Knowledge Hub.

As authors of this book 4impression are uniquely positioned to offer additional personalised training to readers who require more insight into its content. The directors of 4impression, colleagues from their days working for the Jarvis Porter Group, are passionate about print and have a long track record in delivering courses to major packaging users and their supply chains.

John Morton

John has hands-on experience of all the major printing processes and has held operational and technical development roles at director level in the packaging sector. A qualified printer, John's career spans magazine production, commercial print, packaging and label production. Before joining 4impression John was actively involved in the Unilever advanced printing and decoration training courses attended by delegates from operations around the globe.

Robert Shimmin

Robert has held senior marketing and business development positions in the print, packaging and label sector spanning more than 20 years. He is a regular contributor of articles to the print and packaging trade press and has supported initiatives that seek to build awareness of the latest research and innovations emanating from UK universities. In addition to his involvement with 4impression he runs Shimmin Associates, a research and marketing consultancy offering support to both UK and international customers in the label and packaging sector.

Paul Jarvis

Paul Jarvis, formerly chairman of Jarvis Porter Group PLC, oversaw its growth to become one of Europe's leading packaging suppliers with a turnover in excess of £100 million, employing 1,600 people in 7 countries including the United States. Paul was a director and founder member of the Leeds Training and Enterprise Council and represented CCL Label on the main board of FINAT (the world-wide association for self-adhesive labels and related products). Paul provides strategic direction to the packaging and print sectors capitalising on his vast experience and global network of contacts.

www.4impression.com

4 impression

Acknowledgements

Chapter 1

——————

Market overview

——————

Shelf impact and differentiation are critical to the success of any brand. Since the arrival of the self-service retailer it has been recognized that how a product looks on the supermarket shelf will have a direct influence on the purchasing decisions of the consumer.

——————

Brand owners have invested fortunes in their brand image and these are carefully controlled and exploited across the globe. The label or product decoration is a key element in transmitting brand image and is therefore carefully manipulated to maximize shelf impact.

It is difficult to comprehend that until relatively recently there were only two choices when it came to applying a label to a pack, for example, licking a pre-gummed label or getting out the glue pot.

Today however, there are a wide variety of methods used to decorate a pack and it is very easy to make assumptions about the colorful livery that brand owners adopt to make their product attractive to the consumer.

A key aim of this book is to explore most of the commonly used methods of pack decoration, explain why a particular method is used, examine the key characteristics, as well as the processes and technologies involved in producing the finished decorated pack.

DEFINITIONS/SCOPE
This book will cover four key categories of product decoration (see Figure 1.1).

DIRECT DECORATION
Direct decoration involves the application of unsupported liquid ink directly onto the pack or product.

Direct printing using both analogue and increasingly digital methods, is being used for a wide variety of applications and it has evolved for use on

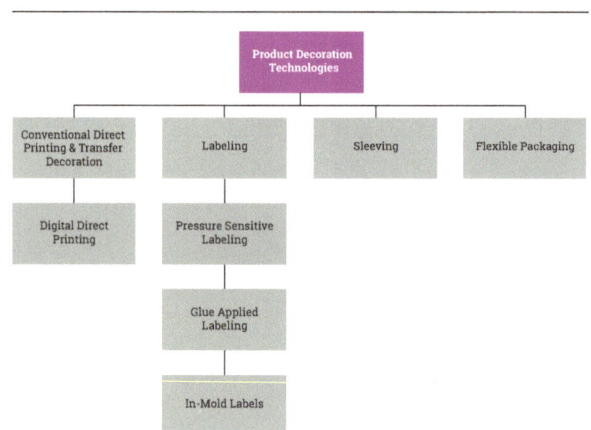

Figure 1.1 Main product decoration technologies

high throughput production lines.

With conventional direct printing, ink can be applied to the pack or packaging material using a variety of printing techniques, but in most cases the ink carrier comes into direct contact with the pack or product and the image is transferred under pressure.

Direct digital printing which is especially suitable for drop on demand inkjet printing is a significant growth area and will be dealt with in Chapter 3. This non-contact process allows ink to be transferred directly onto the product with no contact being made with the pack surface.

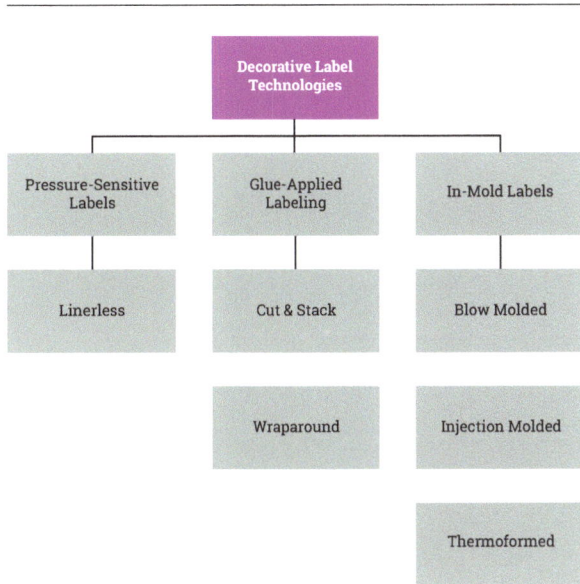

Figure 1.2 Summary of the common decorative label technologies used for pack decoration

DECORATIVE LABEL TECHNOLOGIES

Unlike the direct printing methods, labeling involves the application of an identifying or descriptive marker that is attached to a product or pack usually via an adhesive.

The three main labels systems that will be discussed are pressure sensitive labels, where the adhesive is already coated onto the label, glue-applied labeling where the adhesive has to be applied during the application process and in mold-labels where a label is inserted and attached to the pack whilst it is being manufactured (see Figure 1.2).

LABELING – A HISTORICAL PERSPECTIVE

Attaching a label or descriptive marker to a product is perhaps the earliest form of product decoration. Since their first appearance the role of the printed label has been transformed.

The use of labels today is more diverse and wide ranging than ever before, covering more label processes, technology solutions, materials and requirements than at any time in the previous history of labels. Almost every week sees new label solutions and requirements and even the once paper-only label may now be found in metallic foils, metalized papers and films, in plastics of all kinds, in synthetic papers, metals, fabrics, on rubber, and much more.

From one or two basic mechanical printing processes label technology has advanced to the stage where virtually every printing process – from letterpress to offset, flexo and UV flexo to screen, gravure to hot stamping – can be found. Increasingly label manufacturers and users are producing labels with a variety of digital printing technologies.

In this book we uncover the story behind each labeling system including systems that integrate labels into plastic packing during its manufacturing process (known as in-mold labeling).

The function of the label is simply to say what is inside the pack and in some cases to visually enhance the pack. The label has multiple functions. It needs to provide the consumer or user with all the necessary regulatory information – name of the product, ingredients/contents, manufacturer/supplier contact details, weight/volume, health or safety information, usage instructions, etc., and provide a means to market the product. The information provided on almost all labels today is governed by comprehensive national or international legislative requirements.

It is easy to speculate on when the earliest paper labels may have been used … a piece of papyrus stuck on an earthenware pot 3,000 years ago or a hand-written label on handmade paper made by the Chinese 2,000 years ago. In the fifteenth century paper had become more available, allowing the label to be widely used.

The earliest known 'printed' labels were used during the sixteenth century for bales of cloth. By 1700 printed medicine labels were in use, and possibly wine labels in Italy.

Paper-makers were probably the first to use wrappers with a printed design in the center. It may be these designs were cut out and used as labels.

Until the end of the eighteenth century labels were printed by hand on wooden presses, using handmade paper. In 1798 two inventions led to the proliferation of labels: the paper-making machine, invented in France by Nicolas-Louis Robert; and the principle of

Figure 1.3 An explosion in beverage label printing occurred from the end of the 1800s and throughout the 20th century

Figure 1.4 Heat shrink sleeving provides 360 degree graphics on Smirnoff Sours. Source: CCL Label

lithography, discovered by Alois Senefelder in Bavaria. (see the Label Academy book on Conventional Label Printing Processes).

By the 1830s labels were used on all forms of packaging material and on a wide range of products. The next revolution was to be color printing (Figure 1.3).

Color obviously enhanced the label greatly, but it was expensive to have labels colored by hand. Developments progressed to find an effective, but inexpensive way to print in color and in 1835 George Baxter patented his method of color printing from wood engravings. By the 1850s the process of chromo-lithography (printing on stones using a system of dots and solid areas) had been developed.

For a full historical perspective see the book The History of Labels.

SHRINK SLEEVES

Shrink sleeve labeling is a growing technology, initially used to band two or more packs together for promotional purposes, which has since evolved to encompass the decoration of individual unit packs.

Labels are produced in sleeve form from specially formulated films which are capable of shrinking biaxially around a product.

Shrink sleeve labeling was originally developed in Japan in the 1960s in order to produce shrinkable cap seals for tamper evidence on sake bottles.

Later this was used to band two or more packs together for promotional purposes, before extending to the labeling of individual unit packs in the 1980s.

Sleeves provide 360 degree graphics around a pack offering surface protection from rubbing and a high degree of label security (Figure 1.4).

The 4 main sleeving formats used today will be covered in detail in Chapter 7. These formats are:
- Pre-welded shrink sleeves
- Stretch sleeving
- Reel-fed wraparound sleeving
- Roll-on shrink-on sleeving (ROSO).

FLEXIBLE PACKAGING

The use of pre-printed flexible materials in the primary decoration of products, offers a convenient way of

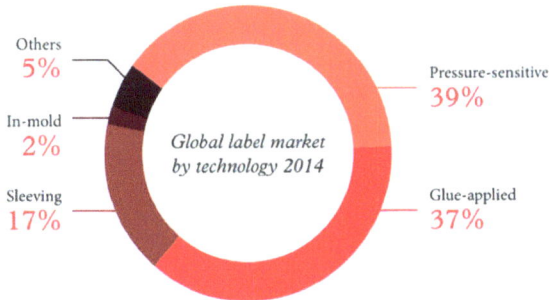

Figure 1.5 Global label market by technology. Source: AWA

Figure 1.6 Printed flexible pouch

integrating packaging and marketing requirement into a single system (with no need for any additional labeling).

The use of printed filmic, paper or foil materials formed into wrappers, packs, pouches, lids, bags or sacks etc to cover or contain a product has many distinct advantages over other decorative packaging.

In effect the wrapper itself acts as both the packaging and the label and therefore requires no further decorative input. It is therefore low cost, light and disposable and is an extremely attractive option for brand owners. Flexible materials can be difficult to convert and there are some issues to consider where the packaging material is in direct contact with say a food product.

Chapter 8 will explore a number of flexible packaging systems in use and their roll as a carrier of product decoration and branding in a variety of market sectors.

Folding cartons which clearly perform a decorative function in many consumer markets such as foods, healthcare and tobacco are not included in the book.

MARKET STRUCTURE & TRENDS

Traditional wet-glue applied and pressure-sensitive (self-adhesive) labels make up approx 76% of all labels used globally, but while the growing usage of plastics in packaging has stimulated the rise of filmic pressure-sensitive labels, it has also opened markets for the newer labeling technologies.

Sleeving technologies (shrink, stretch, wraparound and roll on shrink on) are increasing with a share of 17

percent, in-mold with a 2 percent share, with other labeling technologies making up the balance of the market at 5 percent (see Figure 1.5).

Pressure-sensitive and sleeve labels represent the fastest growing segments of the global label market, with digital label printing now the fastest growing print technology worldwide, with 30 percent of all global label press investment in 2014 being digital presses.

The market for flexibles which accounts for the largest sector of global package printing is expected to achieve growth rates of +4.5 percent annually until 2020.

Rapid growth in recent years has been driven by an array of innovative new origination, print image carrier and press technologies, as well as developments in flexible packaging substrates which have seen quality decorated films become ever more attractive in a consumer-driven world.

Single or re-fill pouches, in particular are being used for an increasing range of products, from liquid detergents to beverages and lubricants – and this is expected to continue to grow rapidly (Figure 1.6).

Each labeling and decoration technology, their role and relevance to different packaging formats will be covered in depth in this book.

Chapter 2

————

Conventional direct printing

————

Direct decoration can be defined as the application of unsupported liquid ink directly onto the pack or product. With direct printing the need for a secondary carrier such as a label is totally eliminated.

————

With conventional direct printing ink can be applied to the pack or packaging material using a variety of techniques such as screen, pad and dry-offset. In most cases the ink carrier comes into direct contact with the pack or product and the image is transferred under pressure.

Direct digital printing is a significant growth area and will be dealt with in Chapter 3. This non-contact process allows ink to be transferred directly onto the product with no contact being made to the pack surface.

Spray coating techniques are also classed as methods of direct decoration and will be covered in this Chapter.

The common conventional direct printing technologies used for the decoration of rigid packaging are summarized in Figure 2.1.

SCREEN PRINTING

Screen printing has been used for point-of-sale, label and industrial printing for many years.

In the industrial field, screen printing is used for clock faces, instrument panels, keyboards, signs, etc.

Screen printing is still a widely used technique for the primary decoration of both glass and rigid plastic containers, although there has been a significant decline in its use at the expense of self-adhesive labels.

Screen printing offers excellent durability, product resistance and strong bold graphics (see Figure 2.2).

Screen printing can be a little messy and often involves flame or corona treatment of the container to facilitate ink and this is why it tends to takes place at the packaging manufacturers, such as the plastic container blow moulders or glass container makers. There are some brand owners however who have direct printing capabilities in-house but they are few and far between.

Most brand owners and contract fillers will use containers that are pre-decorated. This eliminates the relatively slow decoration process from their operations, thereby allowing them to run their filling

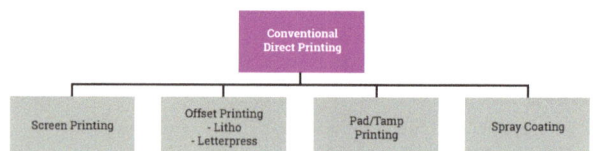

Figure 2.1 Summary of the common conventional direct printing technologies used for pack decoration

Figure 2.2 Examples of direct screen printed containers

lines at faster speeds. However, it eliminates the flexibility to make copy changes, and is expensive if the specific quantity ordered is incorrect.

THE PROCESS

Historically the direct decoration of containers was carried out on semi-automatic single color screen printing equipment.

Typically, the bottle or container is held by the base and the neck and is moved up to the container being printed. Screen printers are equipped with inflation systems to force air into the bottles, giving them the rigidity necessary for the screen printing process. Multi-color printing requires registration of the image. Custom tooling is therefore required for each bottle size and shape to engage the registration feature molded in at the bottom of the bottle. A squeegee blade moves across the screen forcing ink through the open mesh area (the image area) onto the surface creating the printed image. A rack and pinion drive system rotates the bottle at the same speed as the screen moves during the print stroke, ensuring precision printing and multi-color registration.

The bottle is then moved to the next position where a second color can be applied. The equipment has the ability of applying multiple colors at one time, but for more complex designs the bottle can be put through the process more than once.

Normally, print cannot be on the curve of a bottle

shoulder and must usually be around 7mm or so below the curve. Also, print cannot normally be placed on the bottom 20mm of the bottle.

Multi-color direct screen printing was made possible through the development of thermo-plastic (hot-melt) colors. The glass industry in particular was responsible for setting many of the standards for machine construction for container decoration. Handling systems were generally robust to withstand the weight and stress placed on mechanical components by the relatively heavy glass bottles. The same capabilities were soon extended to plastic bottle decoration.

Over time there has been a trend towards the use of fully automatic bottle decoration with multi-color capability. Although solvent based inks are used, the introduction of UV curable ink systems heralded a new era for the direct decoration of bottles.

In the glass industry inks that are applied directly to the surface of a bottle are fired through a Lehr furnace. After firing at temperatures of up to 1180° F the label design is permanently fused to the glass.

Today's automatic screen printing equipment offers multi-function capabilities such as pre-treatment for polyethylene and polypropylene bottles by flame or corona discharge which facilitates ink key. The equipment may also need to interface with other operations such as bottle molding or filling. Equipment must be versatile enough to handle a variety of container shapes and sizes, the registration of requirements of high speed multi-color printing and a wide variety of substrates and inks.

The plastic bottle printer must often inflate his substrate/container to make it rigid enough for printing.

NOTE: The principles of the screen printing process are dealt with in detail in the Label Academy book on Conventional Label Printing Processes.

CHARACTERISTICS OF DIRECT SCREEN PRINTING

The characteristics and benefits of direct screen decoration differ and are dependent on the market and packaging material.

Under-the-sink household products for example tend to use medium density PE bottles for products such as washing up liquids, detergents and bleaches. Featuring simple designs of one or two colors, containers tend to be printed using solvent based inks with infrared/hot air drying. Economy, performance on flexible materials, product resistance and durability are the benefits delivered by the screen process in this sector. Pack size will tend to be large and the printing will often be carried out by the blow-molder in this market.

In the toiletries and cosmetics market the requirement is for a much higher quality of decoration. Typically print will be 3-4 colors onto HDPE containers. In this sector UV ink systems have superseded conventional solvent based systems.

Modular in-line UV printing equipment allows instant drying between stations and can achieve a high output of printed containers per hour.

The advantages of UV screen printing in the personal care market are as follows;
- High visual impact
- Good graphic definition and detail on text
- Instant drying
- Excellent product, durability and performance in wet and humid conditions

It is worth noting that the trend towards PET/PETG containers is offering opportunities to other decorative systems such as self-adhesive labels and sleeving.

In the glass industry high temperature long stove screen inks are used particularly for the decoration of high quality toiletries, cosmetics and spirits brands. Four color process printing on glass is possible when using UV inks whilst other effects can be achieved using precious metals, glow-in the dark inks, and raised lettering.

Screen printing is also used for decorating returnable (or premium brand one-way) glass bottles and more recently PET bottles – including Coca Cola and Pepsi bottles.

For returnable bottles the bottle is screen printed at the point of manufacture with a durable, scuff resistant, image that will endure numerous round trips, including washing operations prior to re-filling. Screen printing of glass or plastic bottles is relatively slow at around 6,000 bottles per hour and could not be carried out on a bottling line. However, if used for decorating returnable bottles, where there may be up to 20 trips per bottle, the process becomes a viable operation and can compete with labeling.

By using different thickness gauge meshes, mesh spaces and inks it is possible to print ink films thicker than any other process, so creating a tactile feel and image to the decoration. With some stencils it is possible to achieve an embossed feel to the decoration.

Four color process printing on glass is possible when using UV inks whilst other effects can be achieved using precious metals, glow-in the dark inks, and raised lettering.

DRY-OFFSET PRINTING

Dry-offset printing (known as offset letterpress) was developed in the early 1960's with the introduction of photopolymer dry offset plates.

The term dry-offset derives from the fact that these new plates could be used on standard litho presses without the use of the water that is needed for conventional litho printing.

With dry-offset the complications of maintaining the ink-water balance associated with conventional lithography are eliminated. With dry-offset however it is not possible to print as small a dot as it is with lithography, so print quality and definition can suffer.

THIN WALL PLASTIC CONTAINERS

Dry-offset letterpress printing is used predominantly for the decoration of thin walled plastic containers in the dairy sectors (fats, ice cream, yogurt, buckets etc).

Offset letterpress uses a photopolymer relief plate (similar to conventional letterpress printing). Unlike conventional letterpress where the ink is transferred directly to the substrate from the plate, in dry offset inks are applied indirectly via a rubber blanket and then onto containers that are held on a rotary mandrel system. On a mandrel system the containers are supported internally to provide a stable surface on which to apply the print.

Due to print quality issues associated with dry-offset there is a trend to in-mold labeling in this sector.

TWO PIECE CANS

Figure 2.3 Dry offset printing on cans using Esko HD letterpress

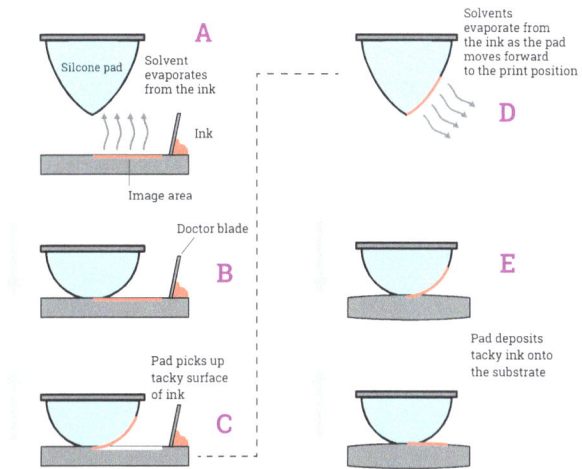

Figure 2.4 Typical pad printing process

The aluminium two-piece can used for soft drinks and beers is printed dry-offset by the container manufacturer.

As with thin walled plastic container decoration, a special plate prints directly onto the blanket of an offset press and the blanket then offsets the image onto the metal.

The inks are only allowed to dry after this transfer.

Dry offset printing for conical/cylindrical packaging, such as beverage and aerosol can decoration, places multiple inks, 'wet-on-wet', on a common blanket and that image is then transferred to the packaging to be decorated. The inks are only allowed to dry after this transfer.

The introduction of High Definition (HD) letterpress with UV exposure and high resolution imaging is overcoming some of the challenges for dry offset printing ie. smooth highlights, clean printing of text; and fine positive and negative linework (Figure 2.3).

PAD PRINTING

Pad printing (also called 'tampography') was developed principally as an economic method of printing awkward, irregular or 'not flat' profiles. It is used for printing on 'difficult to print' products in many industries including medical, automotive, promotional, packaging, apparel and electronic objects, as well as appliances, sports equipment and toys.

Pad printing can also be used to deposit functional materials such as conductive inks, adhesives, dyes and lubricants.

The heart of the process is the printing pad which is made of silicon rubber. It is malleable and therefore able to mould itself to the contours of the area to be printed.

THE PROCESS

Pad printing is a process whereby an image is transferred from an etched plate via a silicone pad, directly onto a component surface.

Physical changes within the ink film allow it to leave the etched image area in favour of adhering to the pad and then it is released from the pad and transferred to the substrate.

The pad printing process is described in Figure 2.4.

A-B The inking slide moves forward, the self setting spatula scoops ink from the reservoir and deposits an even layer across the plate.

The doctor blade scrapes the ink back from the plate leaving a deposit of ink in the etched plate surface. Solvent evaporates from the ink and its surface becomes tacky.

C The pad descends and comes into contact with the ink in the image area where it transfers.

D As the pad travels to the target area solvent evaporates allowing the ink on the pad to become tacky

E The pad descends to the target area which has greater adhesion with the ink than the silicon rubber pad. As the pad lifts from the target area the ink adheres to the substrate.

SPRAY COATING

New developments in the electrostatic spray coating of glass and most recently, PET bottles now look to provide some interesting new possibilities for added-value bottle decoration.

It is also suitable for many other products including cosmetic containers, tableware, lighting glass and decorative glass.

With electrostatic coating technology bottles are automatically sprayed with an electrostatic liquid or powder, giving a range of effects from transparent to opaque, full gloss to matt, silky smooth to stone effects, slippery sleek to rubbery soft, as well as thermochromic or fluorescent images.

Such effects can also provide for a glass etched or frosted look, offering bottles that are colored or textured and which are designed to sell and add value to a product.

The various coating options can all be over-labeled or screen printed.

Coatings are extremely thin and can be recycled with the bottle. Being water-based, the coatings are compatible with organic inks and reduce the amount of overall packaging used.

By combining spray coated barrier coatings with special effect spray finishes and then labeling (which can make frosted or etched effect areas transparent if required), then electrostatic spray coating could provide bottlers with some very exciting added value effects for, say, speciality niche beer brands.

TRANSFER DECORATION

Transfer decoration is a method of product decoration which is an alternative to printing directly onto the container or using conventional labeling systems.

Heat transfer decoration uses a combination of

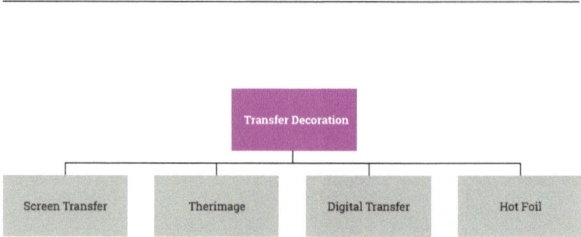

Figure 2.5 Main transfer decoration technologies

heat and pressure to carry an image onto a product via a liner. The result is the look and feel of a label without a label.

Heat is typically required to soften the inks, adhesives and substrate surface to ensure the correct adhesion.

There are a number of heat transfer systems available, the most common of which will be dealt with in this section (Figure 2.5). Each transfer decoration method requires a different type of machine to successfully apply a crisp, quality image to the product or pack.

SCREEN PRINTED TRANSFERS

First introduced in the early 1960's screen printed heat transfers evolved as a decorative method in the plastic molding industry for adding color graphics to consumer electronics and appliances.

Screen printed heat transfers deliver outstanding graphic quality with excellent ink coverage and utilizing custom chemistry designed for the specific material and application.

Screen printed transfers produced in layers (one for each color), that begins with the top color. In some cases a layer of adhesive may be required to adhere to specific substrates.

Printing is typically onto a polyester carrier.

The inks used in heat transfer applications are manufactured to a specific formulation that delivers the performance characteristics required for each application.

An applicator uses a combination of heat, pressure and time (dwell) to transfer an image onto a product via a liner.

WAX RELEASE HEAT TRANSFER LABELING (THERIMAGE®)

The Therimage® (Wax Release) process was first introduced by Dennison Manufacturing Company around 1960.

Therimage or Heat Transfer Labeling (HTL) is a method of product decoration which is an alternative to printing directly onto the container or using conventional labeling systems

The reverse printed graphics are printed onto a special release coated paper. The label construction consists of adhesive lacquer, ink layer, protective lacquer and the release liner.

During the automatic application process only the printed inks or film lacquer transfer to the product.

Therimage will give the same look as the other systems but without using a conventional label substrate and when applied to the container looks as if it has been printed directly on to the container surface and not actually applied as a label.

THE PRINCIPLES OF THE PROCESS

There are two separate elements involved in the printing and application of Therimage/HTL decoration.

Firstly, the printing and laquering process and secondly the transfer and application of the printed layer (label) onto the container.

Therimage labels can be printed using the mainstream conventional printing processes, gravure, flexographic, offset litho and screen. The inks are compatible with all these processes and produce excellent graphics, with the screen process perhaps giving the best result when good opacity is required.

The image is printed in reverse which means that the colors which are normally printed CMYK. i.e. black as the last color would be printed KYMC. i.e. cyan as the last color printed, this means that the 'label' lays upside down on the substrate carrier. After the printing has been completed a lacquer coating is applied which provides the adhesive layer which secures the Therimage label to the container. The carrier substrate is also treated with a special wax

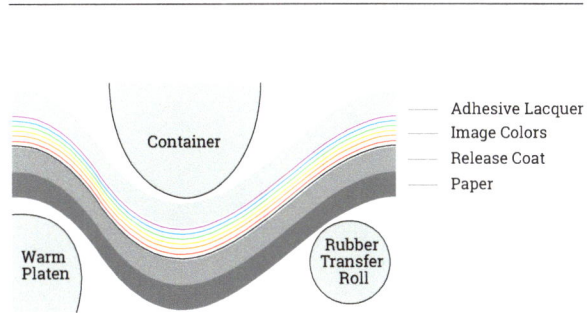

Figure 2.6 Structure of heat transfer label

release coating which allows easy release of the label from the carrier to the container. Figure 2.6 shows the structure of a heat transfer label.

The second stage is the application of the label onto the container, this involves passing the printed web through an applicator machine which transfer the label from the carrier web on to the container (see Figure 2.7).

By the use of pressure and controlled heat, the printed ink layers of the Therimage label are placed in position onto the container giving a permanent and robust decoration.

A pre-heat station is required to allow the inks to soften so that they 'flow' from the carrier to the product.

It is sometimes necessary to flame treat the containers prior to the label being applied, this process oxidizes the surface of the container to increase the surface energy, not unlike corona treating a conventional substrate prior to printing. Recent developments in Therimage ink systems have created inks which do not require the flame process to be used, this 'flameless' technology allows the adhesive to adhere to the container without the use of the flame process.

Advantages
- Therimage labels give excellent graphic reproduction include fine vignettes

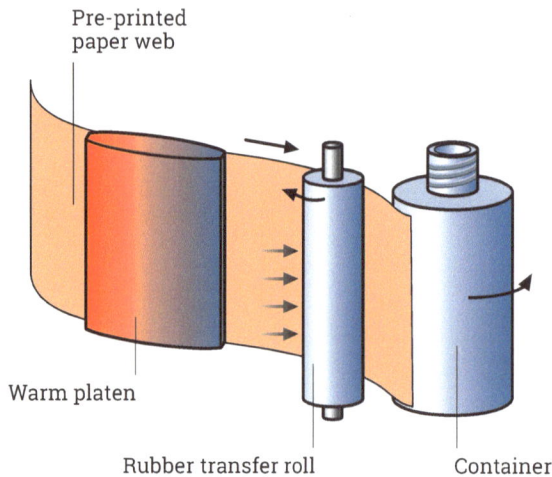

Figure 2.7 Therimage decoration application process

Figure 2.8 CDigital uses Xeikon digital print technology to apply imagery to non flat products

- Low minimum quantities
- Multi-color printing including metallic and acid-etch finishes.

Disadvantages
- Some containers require a flame treatment.

DIGITAL HEAT TRANSFER DECORATION

Patented systems are currently evolving that use digitally printed transfer labels to decorate non-flat or shaped products made from plastics and metals, glass and ceramics (Figure 2.8).

Digital transfers are used primarily for the decoration of consumer goods packed in plastic containers such as seamless tubes, buckets, cartridges, etc.

The system uses dry toner to print digitally onto the underside of patented constructions. The special heat transfer carriers are constructed with a release layer and a lacquer layer. The converter lays down a water-based adhesive on top of the toner in-line with the press. The roll of printed transfers is then fed into an applicator which applies the label onto the container using heat and/or pressure in a process similar to hot foil blocking.

With digital transfers, heat is required to activate

the adhesive coating of the transfer along with the toner printed image.

Digital heat transfers offer excellent, reliable control and the ability to incorporate variable data into a long run.

Containers made from a range of plastics including polypropylene (PP) and high density polyethylene (HDPE) can now be decorated with high impact graphics using this method (Figure 2.9).

SUMMARY

Heat transfer decorating offers a number of benefits over many other decorating processes, especially wet decorating processes. Some of the key advantages are summarized below;
- Dry Process with no wet inks, chemicals etc.
- No discernable label edge
- Multiple colors in a single pass application
- High scratch and chemical resistance
- Permanent with fade resistance
- Custom colors and formulations specific for each application and substrate
- Easy to change graphics
- Decorated plastic containers can be squeezed or deformed without adverse impact on image quality or adhesion.

Figure 2.9 Digital heat transfer onto plastic containers. Digital decorating system jointly developed by Xeikon and Italy-based Moss

HOT FOIL BLOCKING

Hot foil blocking is a common embellishment process used in label manufacturing (see the Label Academy book on Label Embellishments and Special Applications for more details). A similar process can be used for creating decorative metallic effects directly onto to the surface of a product, bottle or jar.

Hot foil blocking is a process that uses metallic dies with a design or lettering etched on to the surface. During the foiling process the colored metallic pigment coating from a ribbon of plastic material known as the 'carrier' is transferred onto the product surface.

The hot foil machine applies pressure and heat to the metallic block, which activates the film. This causes a thin layer of foil to adhere to the surface of the product.

Chapter 3

Direct digital printing

Developments in digital inkjet printing, has the potential to revolutionize the product decoration process. The technology is still in its infancy but it has the potential to meet demand for short-runs, personalization and variable data.

Inkjet printing is a non-impact plate-less process that prints directly from digital data and uses jets of very fine ink droplets fired at the substrate to form the same or variable images onto the product.

Direct inkjet printing can be used on a variety of materials such as PET, wood, paper, metal, foil, glass, and textiles. Inkjet direct decoration has evolved rapidly over-time with cylindrical and conical shapes falling well within the capabilities of the process. Bottles, containers and tubs can now be efficiently decorated using direct inkjet (Figure 3.1).

One packaging sub-sector which has started to adopt direct ink jet printing technology is the printing of cans, bottles and other curved surfaces (Figure 3.2).

The non-contact feature of digital inkjet technology also facilitates printing onto irregular shapes, ridges or grooved areas of a substrate or container which are not suitable for labels or contact printing technologies.

Digital inkjet printing can use a variety of different ink types, including solvent-based, water-based, oil-based and UV inks. UV inks printed onto PET or glass bottles for example offers excellent physical and chemical product resistance, so designs are not damaged by scuffing or exposure to water or other liquids. The UV inks used in the direct print system have a low viscosity and set within a few milliseconds. They have a good opacity, can be overprinted, and have an excellent adhesion to untreated PET bottles. The applied ink is briefly cured by UV light in each printing station. Once the print has been hardened, or

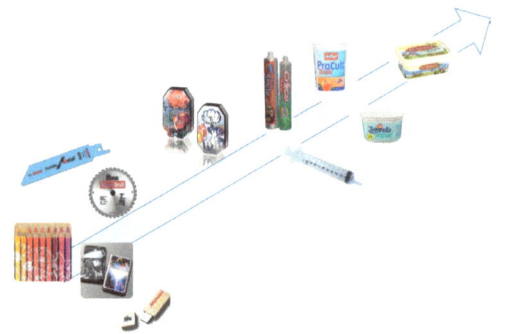

Figure 3.1 Inkjet development in direct product decoration – The transition from flat and semi-flat objects to tubes, cones and tubs (Supplied by Global Inkjet Systems)

Figure 3.2 Examples of direct digital inkjet printing on to PET bottles (Supplied by NMP/KHS)

cured by UV light the labeled bottle is able to withstand all handling in-line, in store and by the consumer.

ADVANTAGES OF DIGITALLY DIRECT PRINTING (DIRECT-TO-SHAPE) OF PET BOTTLES

Digital printing directly onto a (PET) bottle offers significant advantages versus conventional labeling as follows:

1. Supply chain reduction

Where bottles are decorated conventionally using labels, the supply chain can be complex. The label substrate, the adhesives as well as the printing inks, have to be manufactured by different companies and delivered to the label manufacturer. In addition the printing components required for conventional printing (cylinder, plate, screens etc.) have to be prepared. If this is compared to the new process of in-line digital printing directly onto the bottle, the only items that have to be supplied to the bottling line are the printing inks (Figure 3.3).

2. Contactless decoration

As inkjet is a non-impact-printing (NIP) method, the bottle surface is neither compressed nor touched during the decoration process, offering new possibilities for light-weighting, since the bottles do not have to have a highly stable bottle wall to be decorated.

3. Production Flexibility and Personalization

Inkjet printing offers the possibility to instantaneously switch over to different pack designs and there are no minimum order quantities. Each bottle can be decorated differently and mixed packs and bottle personalization become possible.

4. Real 'no label look'

With transparent partially printed wrap around labels, self-adhesive labels and sleeves, brand owners attempt to create a 'no-label look' with the objective of focusing on the product in the bottle and not the label. With the direct-to-shape printing process brand owners can create premium bottle decorations without label materials.

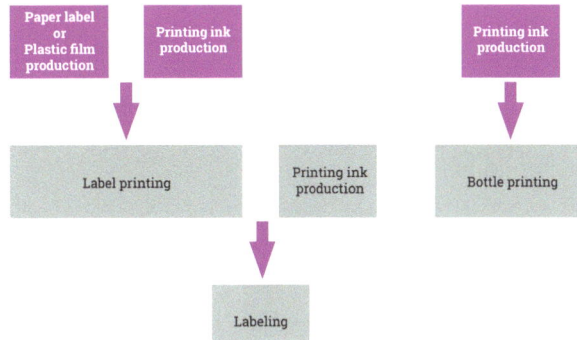

Figure 3.3 Reducing supply chain complexity with digital direct printing

5. Waste reduction

With conventional labeling systems waste is generated for several reasons:
- Minimum order/print quantities
- Obsolete label designs
- Printing and application waste, where reel-fed labels are often wasted in order to avoid stops before and after roll changes.

HIGH SPEED PRODUCTION

Direct inkjet technology, which has been developing since 2006, has made significant progress in the beverage sector.

Today, high speed PET bottling lines run with line speeds of up to 81,000 bottles/hour. Medium speed lines typically fill and label up to 36.000 bottles/hour.

The developers of direct digital print currently use XAAR TF print heads to achieve high resolution graphics (up to 1080 x 1080 pixels).

THE KHS PRODUCTION METHOD

The conventional stop-and-go printing processes where bottles have to pass several static mounted digital print-heads are not suitable for achieving the required line speeds. To increase the print speed, it is essential to transport the bottle whilst printing, which means that the print-heads must be in motion during printing. The KHS production method places print heads on carousels so

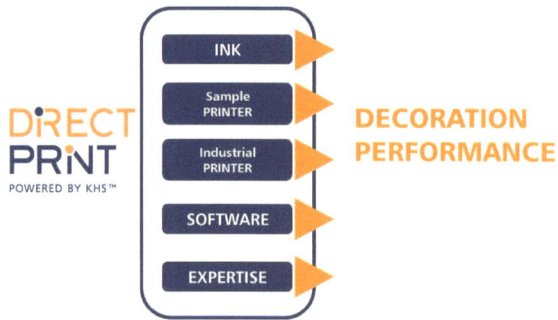

Figure 3.4 System solution to direct print

Figure 3.5 First prototype installed at MARTENS Brewery, Belgium

that they can follow the bottle transport direction whilst printing. In order to make this possible, an ink supply system has been developed to compensate for centrifugal forces and to maintain a constant ink flow through the print-head on a rotating machine.

For speed and handling reasons bottles run in the upright position, which means that the print-heads are positioned vertically (skyscraper mode).

FOOD SAFETY

As ink is applied directly on the bottle wall it is important that no harmful ink substances migrate through the PET bottle wall.

The UV-curable inks used contain photo initiators which form free radicals if they are exposed to UV light. These free radicals initiate the cross-linking of monomers, forming long-chained polymers.

Photo initiators as well as monomers are known to migrate, especially in low viscosity inkjet inks. As PET is not a 100 percent barrier material, special and exclusively (for KHS) developed low migration inkjet inks are used to meet food safety requirements.

RECYCLING

To aid recycling it is crucial that the inks can be removed during different variations of the PET recycling process. Direct Print Powered by KHS™ inks are designed to be removable within the PET recycling process.

NEW CHALLENGES FOR THE BOTTLERS

Digital printing directly onto PET container requires the development of new competencies for the beverage manufacturer. To enable vertical start-up's and smooth market introductions, companies such as NMP Systems provide a system solution which includes: ink, software, sample printer, industrial printer, know-how and service (Figure 3.4).

MACHINE DESIGN

The KHS Direct Print machine has been designed to be modular (Figure 3.5). As with conventional printing machines it is split into color units. In Fig 3.6 one color unit is represented by one printing carousel.

At 12,000 bottles/hour, the system has four print segments (stations) per color carousel. The system ican be upgraded by adding print segments to each carousel, eight print segments total to reach 24,000 bottles/hour, and twelve print segments total to reach 36,000 bottles/hour. This segment structure applies to all processes within the Direct Print industrial printer (bottle pick-up, bottle orientation, printing, curing, print inspection).

PRINT PROCESS

Clean, non-siliconized empty PET bottles enter the print machine via neck handling, transported from the blow molding machine or unscrambler via an air

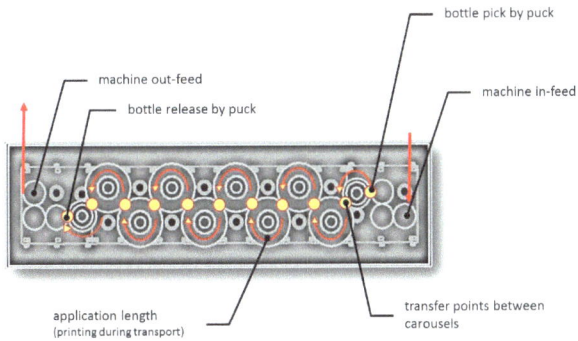

Figure 3.6 Machine design and configuration

Figure 3.7 Illustrates a modular configuration used on KHS digital Direct Print equipment

conveyor. Neck-handling has the advantage in that less format parts are required.

At the in-feed of the direct print industrial printer, bottles are spaced and fixed within bottle carriers ('pucks'). A key role of the pucks is to enable accurate print registration and a consistent flow of bottles through the system. During the whole print process the pucks keep bottles clean and airtight. PET bottle surface pre-treatment (corona, plasma, flame, primer) is not required. Direct print inks are designed to adhere and remain fixed to clean un-treated PET.

INTER-COLOR 'PINNING'

Each print segment is equipped with a single-pass digital print head, combined with a low power UV LED lamp to 'pin' or bond the color on the bottle surface before the next color is applied. This inter-color pinning increases the process stability by freezing the ink droplets after a certain period after printing. Edge sharpness will be increased as the 'wetting' of the droplets on the surface is controlled.

LED CURING

High efficiency UV LED lamps are used for pinning as well as final curing. In contrast to conventional UV bulbs, the LEDs have some significant advantages:

- Mercury free
- On/off switchable within milliseconds

(conventional UV bulbs need a certain warm-up phase, therefore they are normally not switched off during production. The light is only shielded by mechanical shutters)
- the narrow UVA spectrum of the UV LEDs mean they do not produce any ozone
- no infrared light is emitted and therefore the substrate is not subject to as much heat as with conventional bulbs.

PRINT WORKFLOW

There are some innovative developments in the direct print arena that offer users the potential to seamlessly manage their entire print workflow (see Figure 3.8).

Direct Print Powered by KHS™ for example, provides a customer-specific and secure, highly-automated artwork and color management platform which operates in the 'cloud', providing brand owners and their design agencies with a system to prepare artworks, color manage and translate artworks into the digital format for digital printing. With a Direct Print sample printer located either at the brand owner offices or at the agency, brand owners can then print sample (proof) bottles with the correct ink, the correct printing process on the correct bottles – this enables the brand owner to see exactly what they will achieve, prior to sending the same print files via the 'cloud' to the pre-press and workflow of the industrial Direct

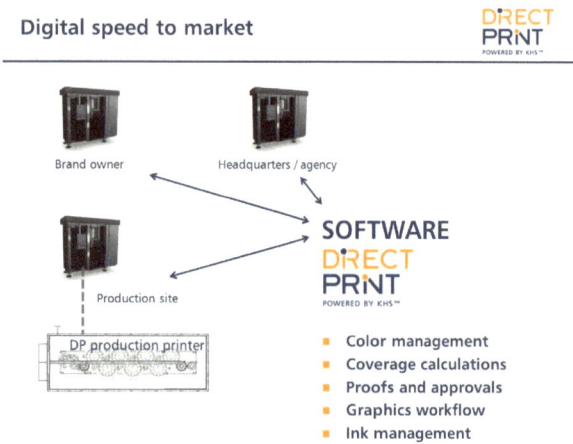

Figure 3.8 Typical direct print workflow

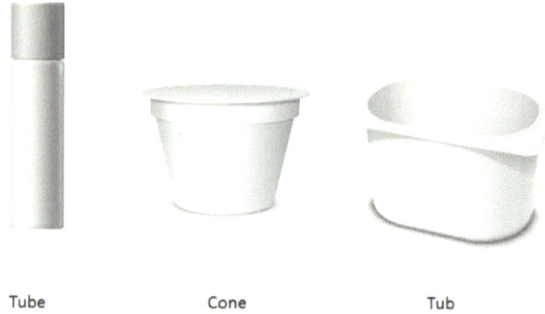

Figure 3.9 Understanding shapes. Supplied by Global Inkjet Systems

Print equipment on-site in the bottling line. Brand owners will be able to efficiently create new decorations, produce print proofs within minutes, and send bottle-to-bottle image variability to their industrial lines in real time – creating unprecedented flexibility and consumer engagement with high resolution white + CMYK decorations.

AN EXPLANATION OF THE INKJET CORRECTION PROCESS

Inkjet print-heads are designed to print onto flat surfaces, so when printing directly onto an object the type of shape to be printed needs to be considered – and in particular the continuity of the object's curves in the direction of print.

A 'continuous' shape is an object whose curvature remains constant in the direction of print – so tubes, cylinders and cones are continuous shapes. In contrast, a 'discontinuous' shape is an object whose curvature changes – an example is a tub (as used for ice cream or butter) as the container is a mixture of flat sides and curved corners making the curvature discontinuous (Figure 3.9).

Today, most of the 'direct to shape' inkjet systems are printing onto containers with a continuous curvature – like tubes and cylinders. If a cylinder is cut

down one side it unfolds and flattens into a simple rectangle or square. This means that the image to be printed onto the tube does not need any image compensation as there is no distortion – the image will wrap around the tube…so essentially the printer is wrapping a 'flat' image around the cylinder.

However, the physical characteristics of the print-head create new challenges. There are three key issues – print-head symmetry, distance between nozzle rows and the number of nozzle rows. The symmetrical orientation of the tube under the print-head is important. For example, if a print-head has two rows of nozzles, the tube should be orientated so that the rows are symmetrical either side of the centre line of the object (Figure 3.10). The narrower the distance between the nozzle rows, the better. Larger print-heads with more rows of nozzles may increase productivity – but the wider the print-head, the harder it is to print onto narrow tubes. At the moment print-heads with two rows of nozzles, a narrow gap between the rows and an overall compact, slim physical size dominate the sector.

Although cones also have a continuous curvature, printing directly onto them is more complex. If one takes a cone and cut it down one side, it unfolds and flattens into a distorted shape that looks like a section of an old vinyl record. Image compensation is required to wrap the image successfully around the cone. There are a number of complexities. If one imagines a

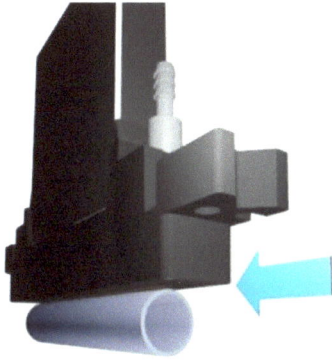

Figure 3.10 Symmetry. Supplied by Global Inkjet Systems

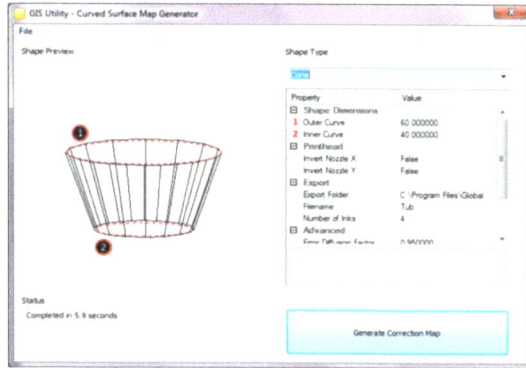

Figure 3.12 Density & screener correction software. Supplied by Global Inkjet Systems

Figure 3.11 Image printed onto a cone with no image compensation. Supplied by Global Inkjet Systems

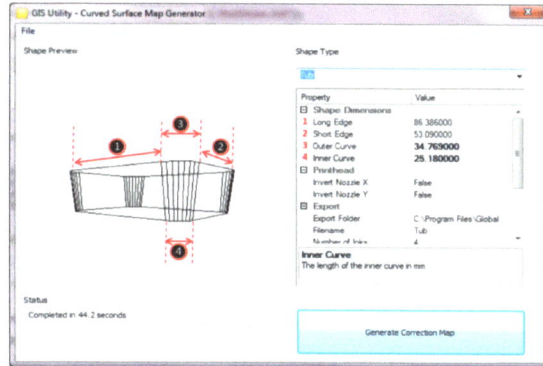

Figure 3.13 Multi-dimensional. Supplied by Global Inkjet Systems

print-head positioned against a conical shape that is rotating – then the resolution changes from one end of the print-head to the other as sections of the cone will be passing underneath the print-head at different speeds, creating higher resolution at one end than the other. Screening the image is also more complex as the image is no longer a square or rectangle. These issues mean that the drops do not land where they should, leading to drop density and dot gain issues - plus nozzle misalignment and time-of-flight differences. Without correcting all these issues a printed image will be distorted, blurred and unacceptable (Figure 3.11). Software is required that corrects drop time-of-flight differences, nozzle misalignment, dot gain and density changes generating a fully compensated image with no screening artefacts (Figure 3.12).

This image compensation software can be taken a step further and used to print onto more complex objects with discontinuous curves such as tubs. The mixture of flat and curved surfaces means that the required corrections change during printing – often pixel to pixel. In this case a multi-dimensional nozzle, density and screener correction technology must be implemented which can be adjusted to each surface type as required (Figure 3.13).

DIRECT DIGITAL INKJET - ADVANTAGES & DISADVANTAGES

Advantages

- Direct printing onto curved or irregular shapes
- Elimination of label materials and adhesives
- No contact with product
- No films, screens, plates or rollers to process or clean
- Variable data printing on demand
- Reduced waste and environmental issues
- Ideal for short run lengths and faster order turnaround with rapid response to market opportunities
- Print on demand with significantly reduced inventory

Disadvantages

- High cost of inks
- High capital cost of equipment
- Low opacity and low coating weights
- Potential problems relating to recyclability of printed containers

CONCLUSION

Developments in direct digital printing has the potential to revolutionize manufacturing processes in the packaging sector.

Already making inroads into mainstream container decoration, direct-to-shape printing is set to develop rapidly as the technology evolves, thereby making it suitable for a much wider range of container shapes and curved surfaces.

Digital processes deliver many advantages in the form of production flexibility and as the cost inks reduce this process will become even more appealing as a decoration system.

Chapter 4

———

Pressure-sensitive labels

———

A self-adhesive label is manufactured using a material coated with pressure-sensitive (permanently tacky) adhesive. The adhesive typically has been protected by a covering of silicone coated material which acts as a carrier up to the point of application

———

It was in 1935 when Stanton Avery of Los Angeles first manufactured the self-adhesive label, that labeling began to evolve from traditional wet-glue methods.

He made the first labels that could be applied without moistening or glue. His first product was the Kum Kleen self-adhesive label offered in only one color: white (Figure 4.1). When customers asked to have their names on the labels, he added print, using a hand-cranked printing press. Self-adhesive labeling usage has expanded rapidly since those early days.

In the beginning, the face of the label consisted of plain, white paper. Today the label face substrate may be made of fabric, plastic, or metallic films and it might carry brightly colored designs and pictures.

The self-adhesive system offers a labeling flexibility that wet-glue cannot provide. The lower cost of change parts needed for the applicator is much lower that wet-glue systems, the set-up time for self-adhesive application is much simpler and operates much cleaner.

The modern combination 'platform press' used in the self-adhesive industry will produce high quality graphics and embellishments in one pass on the press, giving the facility to produce the printed, embellished and converted labels in a much reduced manufacturing window, compared to wet-glue labels.

The use of plastics for packaging, which has shown a faster growth rate than any other packaging material, has particularly helped boost pressure-sensitive labeling. This especially applies to the

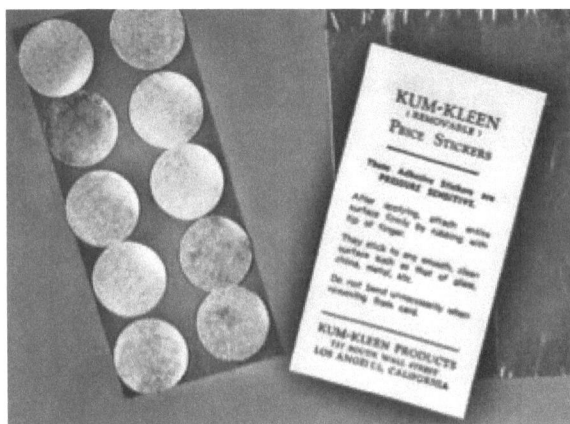

Figure 4.1 Early Kum-Kleen pressure sensitive labels, first produced in Europe by Sessions of York in 1938

shorter-run and luxury end of the market. Typical sectors include toiletries and cosmetics, household products, motor oils and pharmaceutical goods.

Glass containers for foods, drinks and dairy products have historically been labeled with wet-glue labels, or are printed direct. Nevertheless, certain designer products, such as drinks and beers now appear with pressure-sensitive labels. The usage of 'a no-label look' filmic label is of particular note here (Figure 4.2).

Figure 4.2 Growth in use of filmic labels with high decorative content for new premium beer sectors

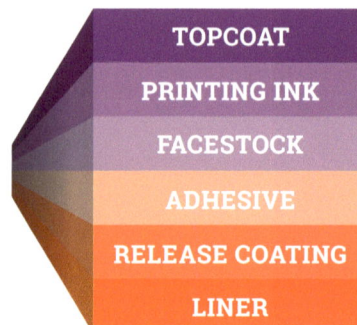

Figure 4.3 Components of a pressure-sensitive label

METHOD OF MANUFACTURE

Pressure-sensitive labels, also known as self-adhesive labels, have been considerably refined over the years.

It is essentially a sandwich of substrates (Figure 4.3). The first part of the sandwich is the face material, which becomes the finished printed label. The reverse of the face material is coated with a pressure sensitive adhesive and the face paper is then laminated onto the liner substrate which is coated with a silicone layer.

A self-adhesive label must have both the ability to 'hold' and the ability to 'release'. It is a contradictory requirement that it must stick to different types of surfaces under varying conditions, yet release easily from its backing paper.

This release coating permits easy removal of both the waste matrix and the label from the release liner, which becomes the carrier for the converted labels. Printing takes place on the face stock and is then die-cut to a shape and supplied in reels.

The face stock can be made from a wide range of substrates, with paper stock or filmic stock being the most widely used. These can be surface coated or uncoated depending on the properties required for the label surface.

Pressure-sensitive adhesives offer an extensive range of adhesive properties. The adhesive/labels are used in a wide range of conditions which demand very high performance levels. These conditions will include high and low temperatures, wet and dry environments and different product surfaces.

MANUFACTURING PRESSURE-SENSITIVE (PS) SUBSTRATES

The manufacture of PS substrates is primarily a coating and laminating process.

The adhesive coating and the laminating of the substrates is done on a coating machine as an in-line operation. A release coating is applied to the liner stock and then dried in-line. The face stock may also require a primer coating and this is also dried in-line. The adhesive coating can be applied directly to the reverse of the face-stock, or by transfer from the liner.

After the coating and drying operations has been completed the two separate substrates (the face-stock and liner) are laminated together to form the pressure sensitive substrate.

The method of PS manufacture is illustrated in Figure 4.4.

There are two methods of applying the adhesive coating to the reverse of the face-stock. The first and most widely used is the direct coating method in which the adhesive is coated directly onto the reverse of the face-stock. The second is transfer coating when the adhesive is coated onto a release liner and then transferred to the reverse of the face-stock after the lamination process.

There are a number of different coating methods

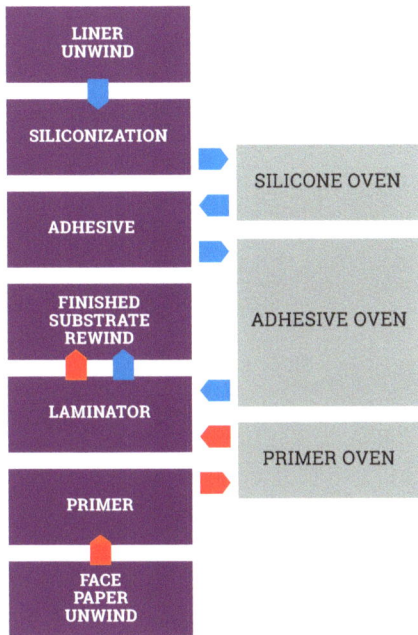

Figure 4.4 Typical method of pressure-sensitive manufacture

Three-roll reverse-roll coater

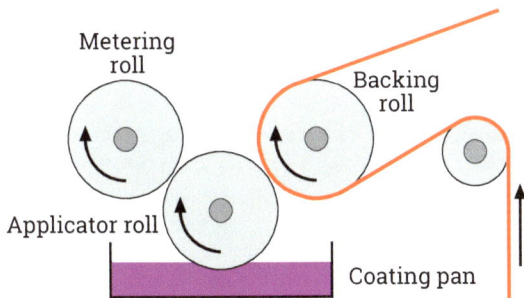

Figure 4.5 Adhesive application using reverse roll coating

used for applying the adhesives.

The thickness of the adhesive is an important factor and must be established before the adhesive is applied.

The coating weight is important for two reasons:-
- It has to be even and consistent to provide the correct adhesion between the liner and the face stock and the label and the product container being labeled
- To apply the optimum/minimum adhesive coat to keep usage to a minimum

Typical coating weights would be in the range of 5-50 gsm for rubber based and aqueous based adhesives

The thickness of the adhesive coating is determined by the following factors:
- The type of face stock to be coated (absorbent – non-absorbent)
- The adhesive system to be used (solvent, emulsion, hot melt)
- The required thickness of the adhesive 0.8 - 5.0+ mils

These factors will determine if the adhesive should be a direct application to the face-stock or transfer coated to the liner.

When this information has been established it is easier to identify the most suitable method of applying the adhesive coating.

APPLYING THE ADHESIVE

There are several different methods of applying the adhesive. It is the type of adhesive that governs the method which is used for the adhesive application.

A typical coating system for aqueous and rubber adhesive coatings would be a Mayer Bar system or a metered system such as a gravure process application.

MAYER BAR COATING METHOD

The Mayer Bar method of application is commonly used to apply low-viscosity pressure-sensitive adhesives (PSAs).

This coating system uses a wired rod to apply the adhesive and is best suited to adhesive types with good flow characteristics.

REVERSE - ROLL COATING

Reverse roll coating (also known as roll-to-roll coating) differs from other coating methods by having two reverse-running nips.

The metering roll and the applicator roll contra-

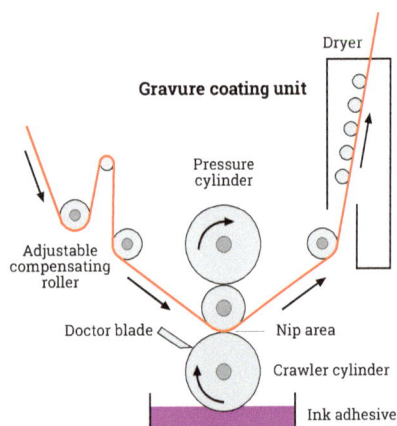

Figure 4.6 A typical gravure coating unit

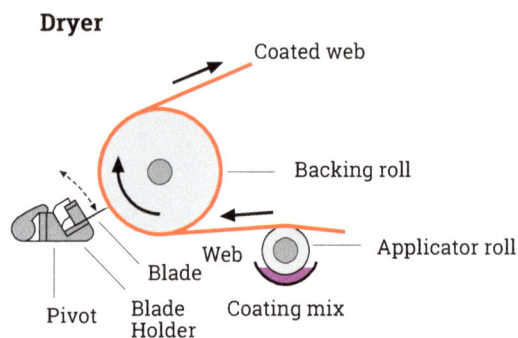

Figure 4.7 The knife coating process

rotate and have a fixed gap between the two rollers. The surface of the applicator roll carries an excess of adhesive coating and the thickness of the adhesive coat is established by the nip distance between the two rollers (Figure 4.5).

The meter roller (contra-rotating) wipes the surplus adhesive to leave a precise amount on the applicator roller. The applicator roll running in the opposite direction to the substrate, wipes the adhesive coating onto the substrate.

GRAVURE COATING

The gravure method of adhesive coating gives an accurate coat weight without any restrictions on the running speeds of the applicator. A low-viscosity adhesive requiring a low coating weight can therefore be applied at speed.

The coating weight (the amount of adhesive being applied to the substrate) is controlled by the screen cells engraved on the gravure cylinder. The depth and size of the cells can be engraved to a specified size. Smaller shallower cells deliver a lighter coating than a deeper larger cell formation, which delivers a heavier coating weight of adhesive.

Gravure applied coating give a very consistent and evenly coated result, suitable for coating clear filmic substrates used in the pressure sensitive industry.

Figure 4.6 shows the principle of the gravure process with the engraved gravure cylinder running in the tray which holds the viscose adhesive. The doctor blade wipes the cylinder leaving the adhesive in the cells. The adhesive is then transferred onto the substrate which then enters the drying section.

For more detailed info on the gravure process see the Label Academy book on Conventional Printing Processes.

KNIFE COATING

In the knife coating process the coating is applied to the substrate directly via a holding reservoir or an applicator roller (Figure 4.7). The knife which controls the thickness of the adhesive, can be a steel blade or alternatively an air knife. The excess adhesive is removed by the knife which is set to a predetermined height or air pressure. This controls the thickness of the adhesive coating by wiping the surface of the adhesive, giving an evenly applied coating at the required weight.

HOT-MELT COATING

The hot-melt coating process is a system of applying an adhesive with a 100 percent solids content of wax and polymer resins. The adhesive is heated to a fluid state and then applied to the substrate by an

PRESSURE SENSITIVE ADHESIVES

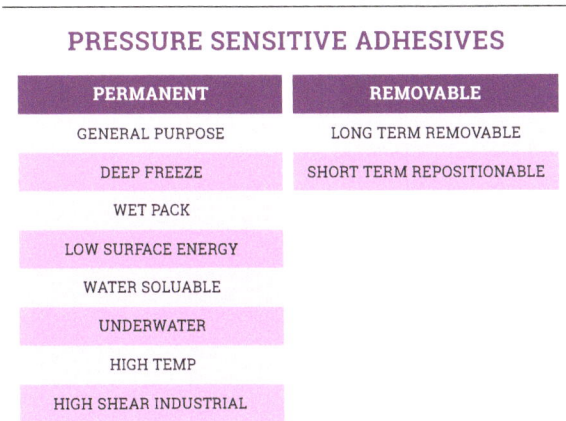

PERMANENT	REMOVABLE
GENERAL PURPOSE	LONG TERM REMOVABLE
DEEP FREEZE	SHORT TERM REPOSITIONABLE
WET PACK	
LOW SURFACE ENERGY	
WATER SOLUABLE	
UNDERWATER	
HIGH TEMP	
HIGH SHEAR INDUSTRIAL	

Figure 4.8 Summary of pressure-sensitive adhesive types

applicator roll, the gravure process or extrusion process. After the adhesive has been applied it is cooled using a chilled roller or chilled air system.

Temperature control is a key factor when using hot-melt adhesives. The temperature controls the adhesive viscosity and the thickness of the adhesive film. This in turn will affect the tack value and the speed of setting i.e. the thicker the adhesive coating the longer it will take to set.

HOT-MELT ADHESIVES

Hot-melt adhesives are 100 percent solids which are supplied in dry form and which melt at temperatures between 266 and 320 degrees F, (130 and 160 degrees C). After the coating process the adhesive coating dries at room temperature, leaving a tacky pressure-sensitive surface.

After the hot-melt coating has been applied, the backing liner is applied to the adhesive coated substrate, thereby laminating the substrate and backing liner together to form the self-adhesive material. The coated material is then passed over a chilled roller to complete the process.

Hot-melt adhesives are used for many adhesive applications, where their immediate speed of setting/bonding is ideal for fast production. It is often used for more difficult labeling applications where emulsion

adhesives can be problematic.

With the hot-melt process higher adhesive coat weights improve adhesion to rough surfaces, but hot-melt self-adhesive formulations are generally unsuitable for labeling plasticized surfaces. The plasticizer will migrate into the adhesive and degrade it, so causing adhesive stringing and strike through of label papers, which may result in bond failure.

Hot-melt adhesives are popular because they are invisible on glass, PET and other plastic containers. They are also suitable for metal applications. Their high tack characteristics make them suitable for difficult applications such as peelable, chill, permanent and deep-freeze label applications.

CONSIDERATIONS WHEN SELECTING PS ADHESIVES

When selecting an adhesive, the type of application and end use of the product to which the label is to be attached, will often determine the type of adhesive to be used.

Key considerations are:

- Is the label to be permanent or removable?
- Storage environment including the range of temperatures, moisture and their potential effect on the label
- The surface to which the label will be applied i.e. smooth/rough, hot, cold, frosted, wet, dry, paper, plastic or metal
- Will label be used for direct contact with food?
- Is there a medical application involving a sterilization process or skin contact?
- Is the label to be used for children's toys?

TYPES OF ADHESIVE

There are a number of variables involved in the manufacture of pressure-sensitive adhesive products (PSA).

Although the process of coating and laminating is uncomplicated, not all PSAs can be coated using every type of coating system.

PSAs use differing adhesive technologies and each of these technologies have varying ranges of viscosity, solids content and coating weights, which may be required for specific applications.

Each of these items can affect the adhesive-coating. These variations will affect the adhesive properties and the choice of the right adhesive and method of application is very important. Figure 4.8 summarizes the main types of PSAs.

MAIN TYPES OF PRESSURE SENSITIVE ADHESIVES

The main adhesive formulations used for pressure sensitive adhesives are as follows;

- Acrylic
- Emulsion acrylics (water-based)
- Solution acrylics (organic solvent-based)
- Radiation curable (UV and EB) – 100 percent solids
- Rubber
- Solvent rubber
- Hot melt rubber (100 percent solids)
- Radiation curable (UV and EB-electron beam) - 100 percent solids
- Silicone adhesives

BASE COATING AND TOP COATING OF SUBSTRATES

There are other processes that are used to optimize the performance of materials. Base coating is used to create a base layer to assist the printing and embellishment processes by improving the ink key and ink lay, and by producing a smoother substrate surface, particularly when the foiling process is used.

The method of application for this type of coating is predominantly via the flexo process, although the screen and the gravure processes are also used. Roller coating too is also used as a base coating process.

The printing units are generally positioned immediately after the unwind unit of the press and can be used to apply an overall coat or alternatively a spot coating.

Top coating or over varnishing of the face substrate is widely used in the self-adhesive label market. The coating is applied as a protective layer after the printing process has taken place giving protection from abrasion and increasing the product resistance. It is also used to produce an alternative finish to the label and enhance the shelf display qualities of the product.

Top coating is also used to aid ink adhesion when secondary overprinting is required, for example date or batch coding data.

Top coated UV over varnishes are typically used to assist product detection via the label applicator or to improve filling line performance.

CORONA DISCHARGE

Corona treatment is another process often used to improve the basis for adhesion of applied printing inks, adhesives, lacquers, etc.

To obtain good adhesion it is necessary to increase the surface energy of the substrate. The surface of film or other material to be treated is bombarded with electrons to increase the surface wettability of the substrate (generally filmic).

Corona discharge is often used on polymer-based substrates that have low surface energy leading to poor adhesion of inks, glues and coatings.

A full explanation of corona treatment can be found in the Label Academy book on Conventional Printing Processes.

MATERIAL SELECTION

Every label application must be carefully considered to ensure that the substrates selected can meet the expected performance criteria.

The following issues will impact on substrate selection;

- The ability of the substrate to withstand the printing process to be used
- The level of absorbency required i.e. the ability of a label substrate to retain liquids, moisture and inks
- The required rub resistance of the label surface
- The required material surface characteristics i.e. smoothness/roughness
- The chemically compatibility of the material to cater for inks, varnishes, solvents and the contents of the products the label may come into contact with
- Stability in different end-use environments and during application and handling

SUBSTRATES USED IN THE MANUFACTURE OF PRESSURE SENSITIVE LABELS

A wide range of paper and filmic materials are used in the manufacture of pressure-sensitive labels.

Figure 4.9 Example of filmic label used in the Health and Beauty sector

PAPER SELF-ADHESIVE SUBSTRATES
The main types of paper substrate used in the label market give excellent anti-fungal and wet strength features and fall into the following categories;

Coated Paper – Gloss and Semi-Gloss
* Matt coated
* Semi-gloss
* Gloss
* Cast coated

Uncoated Papers – matt finish
* Litho
* Vellum
* Textured paper - gives a 'laid' traditional appearance
* Metallics - highly reflective surface – supplied in a range of colors and tints and holographic effects

FILMIC AND PLASTIC SELF-ADHESIVE SUBSTRATES
Whilst paper face materials continue to be the most common face-stock used in the label industry there has been an increasing rise in the use of filmics (Figure 4.9).

These substrates are available in a wide variety of constructions, such as co-extruded, blown and engineered films, all with a wide range of functional properties.

Some of the advantages and disadvantages of filmic substrates are highlighted below.
Advantages
* Can offer transparent features including a 'no-label' look when used on clear containers
* Durable and can offer conformable, 'squeezable' properties.
* Resistance to chemicals, grease, water, moisture etc
* Plastic labels can be compatible with the packaging material being used and this can aid recyclability
* High gloss finish without over laminating or varnishing
Disadvantages
* Can be more difficult to print and convert
* They are often more expensive than paper substrates

Filmic substrates fall into four main groups which form the majority of the film face-stock used today. These are PE, PP, PET, and PVC.

POLYETHYLENE – PE
Low density polyethylene has the highest share of the filmic facestock market and is widely used in the primary self-adhesive labeling market, particularly in health and beauty and the household chemical markets.

There are two types of PE film which are available; Blown PE and Cast PE films.
The key characteristic of polyethylene labels are;
* Excellent tear resistance
* Minimal shrinkage
* Good printability
* Low stiffness
* Good squeeze-ability
* Good resistance to moisture and chemicals
* Environmentally friendly
* Low transparency
* Can be prone to stretching

POLYPROPYLENE – PP
Polypropylene films offer cost-effective and high-yield face-stock and are steadily challenging other films for primary product labeling and variable information printing.

An inherent advantage of polyolefin films over

other film types is their lower density range, facilitating removal and recovery in recycling operations with non-polyolefin containers and in their obvious compatibility with polyolefin containers.

Other key characteristic of polypropylene films labels are as follows;

- PP films may be stretched in one direction (OPP) or two directions (BOPP), or co-extruded usually in 3 layers
- High resistance to tearing and excellent dispensability
- Good die-cutting and printability
- Film flatness and resistant to moisture, abrasion, chemicals
- Good dimensional stability (OPP/BOPP) and excellent conformability
- High clarity and low cost

POLYESTER (PET)

PET films are preferred where durability, higher temperatures and enhanced chemical resistance are required. The major applications are in consumer durables and in the automotive sector.
The key characteristic of PET labels are;

- High durability and clarity and therefore Ideal for clear-on-clear/no-label look applications
- Good resistance to heat, stretch, tearing, chemicals, moisture
- Good dimensional stability and dispensability
- High cost

MANUFACTURING BLOWN AND CAST FILMS

Filmic substrates are manufactured by extruding liquid polymers. This process involves feeding polymer materials from a hopper into an extruder unit in pellet, powder or granule form. The material is heated and the liquefied polymer is injected into a die and by using air directed through the die a bubble, is formed at the desired thickness of film.

The manufacture of cast film differs from blown film manufacturing as the liquid polymer is passed through a flat die which produces flat film.

After leaving the flat die, the continuous sheet of liquid film is cooled which freezes the film, the edges of the film are trimmed and the film is wound into reels.

Cast films can be co-extruded in multi-layers

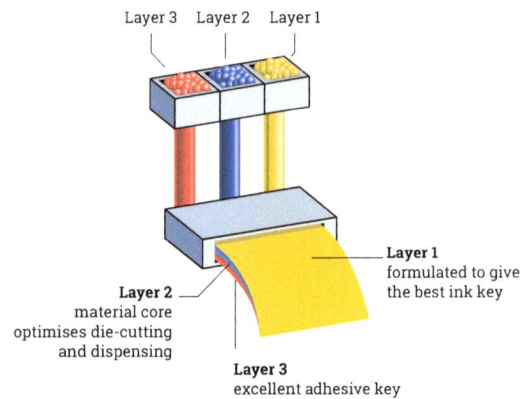

Figure 4.10 Cast films can be co-extruded in multi-layers

(usually three) with each layer optimized for different characteristics (Figure 4.10).

RELEASE LINERS

The release liner (also known as the backing paper or carrier) forms the base of the self-adhesive substrate sandwich. The surface of the liner material is coated with silicone which allows the label face substrate to be easily and accurately dispensed on the labeling line.

Release liners are produced using both paper and filmic materials.

Super-calendared papers offer a robust and consistent surface which is an important pre-requisite for accurate and consistent die-cutting. They also offer good transparency which allows easy label edge detection during application. There are three main types of paper liners;

- Glassine
- Kraft - super calendared
- Paper with thin laminated PE film

FILMIC LINERS

Filmic liners offer a number of advantages over paper liners.

They are light and thin and because of their very smooth surface give an excellent wet out for the adhesive layer. This property enhances the adhesive transparency necessary for missing label identification.

The high tensile strength of filmic liners makes them ideal for high speed label application by removing the possibility of web breaks on the filling and label applicator lines.

There are two main types of filmic liners;

- Polyester
- Polypropylene

THE PRINTING PROCESSES USED IN THE MANUFACTURE OF SELF-ADHESIVE LABELS

Self-adhesive label substrates are printable by all the conventional printing processes, and increasingly by digital methods.

Many paper and filmic grades are also printable by dot-matrix, thermal, laser and inkjet printers for encoding, numbering and personalization.

Details of the conventional and digital printing processes used can be found in the Label Academy books on the subject.

Roll-label converters use narrow-web presses with integral die-cutting and waste matrix rewinders to print and convert self-adhesive materials.

Self-adhesive labels can be over-varnished, film laminated, hot-foil stamped and embossed.

The wide variety of decorative and surface embellishments techniques can be explored in the Label Academy book Label Embellishments and Special Applications.

SELF-ADHESIVE LABEL CONVERSION

In label production, converting covers any process performed to manufacture a complete label from a raw material or an unfinished material. The process of transforming rolls of self-adhesive material into labels on a release liner (carrier), presented in rolls or sheets so as to enable end users to apply them to products, packages or surfaces. This process includes slitting, die-cutting and matrix stripping operations.

A summary of the main conversion processes are included here but for a more detailed explanation see the Label Academy book on Die-cutting and Tooling.

DIE CUTTING

In the die-cutting process the face material is profile cut using a cutting tool to form the label shape. The liner or backing material is kept intact to protect the

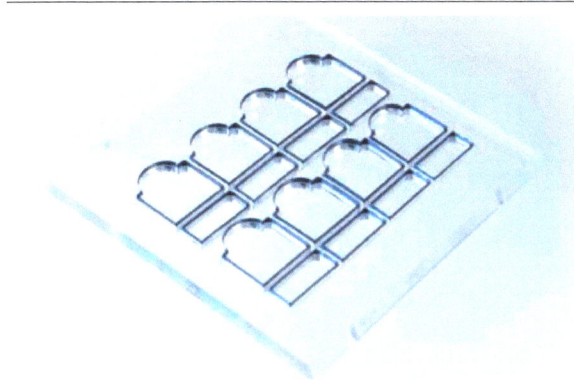

Figure 4.11 Flatbed die-cutter. Source Wink

Figure 4.12 Magnetic cylinder & flexible die. Source: Rotometrics

tacky adhesive and provide support to the cut-out labels, through to the moment of application.

The die-cutting process can be a flatbed or rotary system dependent on the design of the press. Fastest output is achieved when both printing and cutting operations are rotary, as this permits a continuous pass of the laminate through the press. With flat die-cutting the laminate has to either stop and start or reverse during its pass through the press, slowing the overall throughput.

Cutting tools come in a number of forms. Flat tools are normally made up from bent steel rule some 0.4 mm in width and 12 mm in height (Figure 4.11).

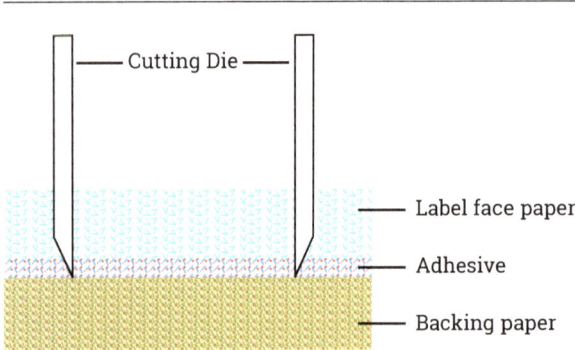

Figure 4.13 The die-cutter needs to cut through the label face material and adhesive, but not the silicone coating of the backing liner

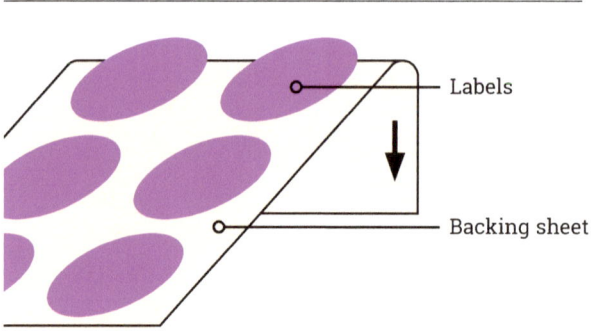

Figure 4.15 Pressure-sensitive labels dispensing as the backing sheet is peeled back under tension

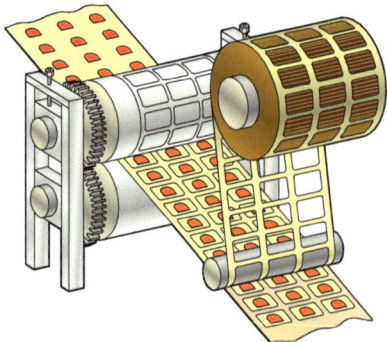

Figure 4.14 Typical rotary die-cutting and matrix rewinding unit

Rotary tools may be formed from a solid bar of steel around which the cutting profiles are etched or engraved, leaving them projecting from the surface.

Laser cutting systems are also used for the for manufacture of flexible die-cutting shims made of sheet-steel some 0.50 mm thick, these flexible cutting shims are mounted on a magnetic cylinder and give extremely accurate die-cutting and manufacturing consistency (Figure 4.12).

The precision and care required for the die-cutting operation is very important. The cutting tool must be in good condition and be located in a die-cut unit that is accurately engineered and equipped with an anvil roller that runs true and is undamaged. The unit must also have the facility for fine adjustments to be made to the pressure system used to control the die tool and anvil roller pressures used for rotary die-cutting.

The depth and consistency of the cut is very important and Figure 4.13 shows the correct depth of cut when converting self-adhesive substrates.

WASTE STRIPPING

In order to achieve maximum economy and depending on the size of label and width of press, the labels may be printed more than one across the width of the press. Waste margins between and around each label may then be stripped away. This process is done immediately after the die-cutting operation and involves removing the waste matrix as a continuous skeleton, the waste is reeled up as the press runs (Figure 4.14).

METHOD OF SELF-ADHESIVE LABEL APPLICATION

A comprehensive explanation of label dispensing and application technology can be found within the Label Academy book on the subject but a summary of the self-adhesive label application process is featured below.

The application of self-adhesive labels can be performed using a table-top, stand-alone, or in-line device/machine which is used to automatically apply pressure-sensitive labels from a roll of labels onto a

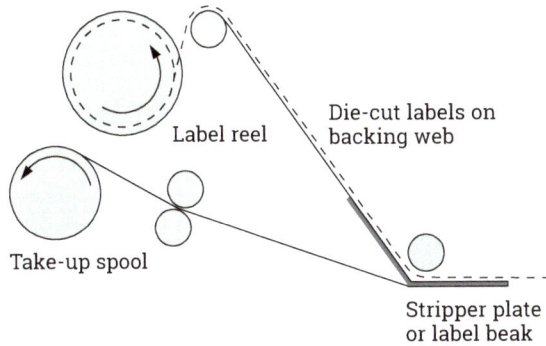

Figure 4.16 Principle of applying self-adhesive labels using a stripper plate or beak

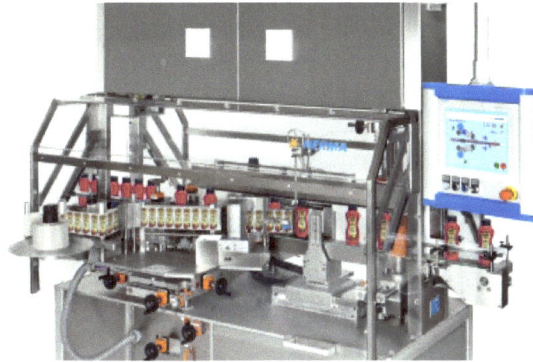

Figure 4.18 Herma 362M labeling system

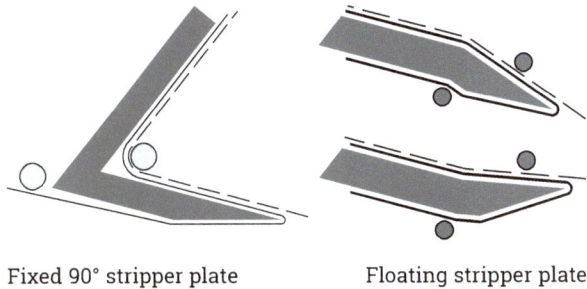

Fixed 90° stripper plate   Floating stripper plate

Figure 4.17 Different configurations used for stripper plates/beaks

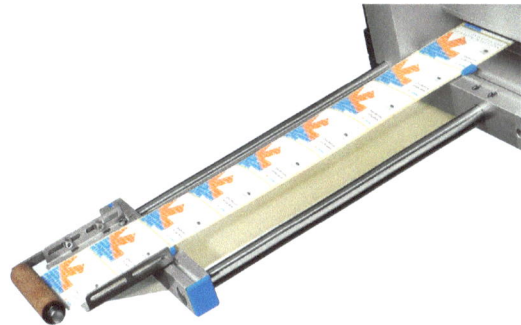

Figure 4.19 Use of a roller to press the label to the product or pack. Source: Herma

container or product.

There are many different types of pressure-sensitive applicators, but they all have one thing in common: a means of peeling the release liner or backing away from the labels. This is usually accomplished by unwinding a reel of die-cut labels and then pulling under tension around a stripper plate or beak. As the backing is taken around the sharp angle of the plate the front edge of the label is peeled away (Figure 4.15 and Figure 4.16).

Stripper plates may be found in a variety of configurations – fixed plate, floating plate, pneumatic, etc. (Figure 4.17).

Rigid stripper plates for example, can be used for very uniform products, while spring biased plates may be recommended for certain side labeling applications.

Floating stripper plates, either solenoid or pneumatically operated, may be used to place labels accurately into recessed areas on a pack at a required point. Powered stripper plates need to be accurately controlled by, say, electronic timers, so that the label is pre-dispensed or maintained in contact with the pack for the correct period.

Once the labels have been detached from the backing there are various ways of feeding them forward and pressing (applying) them on to the pack, container or product to be labeled in the correct position. An example of a self-adhesive label applicator can be seen in Figure 4.18.

The effectiveness of the labeling operation is determined by the uniform application of pressure to the label in order to ensure a positive contact of the adhesive with the surface of the product being labeled.

In their simplest form various rubber or foam rollers and/or drum application devices are used to fix the label to the product, perhaps also supplemented by brushes (Figure 4.19).

Other methods of uniformly applying pressure to the label to ensure that it sticks with a positive contact to the product or container include pressure pads, pneumatic plungers, air jets, knurled rings or powered rotating brushes. Alternatively, the label may be held in position on a vacuum box or drum and released onto the container when it is in the correct position. Once the label has been applied the backing paper is rewound on to a take-up spool.

Roller and/or drum application devices are one of the most common methods of affixing pressure-sensitive labels to products or containers. Ideal for flat or cylindrical packages, they may also be used to label concave and convex surfaces. By employing two application heads it is possible to apply labels to the top and bottom of objects simultaneously, with only one pass through the machine. By setting the applicator heads horizontally it then becomes possible to apply labels to the sides of the objects.

With two units facing each other it is possible to label both sides of the object simultaneously. With some systems it is feasible to apply labels onto the corners of boxes, such as medicines or confectionary box sealing. A variation of conventional roller application used on some machines is the use of rollers made up of knurled aluminum rings, so allowing packs with uneven or flexible surfaces to be handled. Powered rotating brushes, sometimes profiled to the corner of the product, may also be used to press the label into difficult areas.

In some methods of label application a pad or plunger is used to fix the labels in position. These methods are often referred to as Tamp-on applications. Here the dispensed labels are sucked

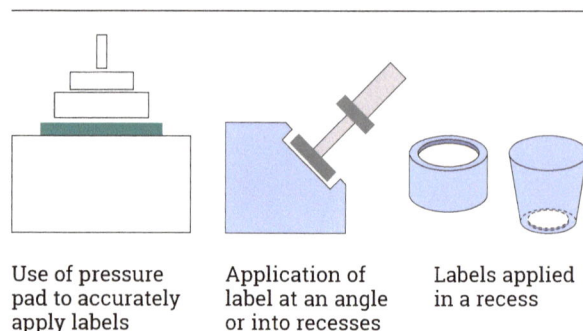

| Use of pressure pad to accurately apply labels | Application of label at an angle or into recesses | Labels applied in a recess |

Figure 4.20 Examples of tamp-on label application

onto a vacuum pad, perhaps on the end of a plunger, and then applied to the product when it is in the correct position by releasing the suction. The result is a highly accurate labeling system (Figure 4.20).

With such systems it is possible to select any surface, direction or angle, up or down, simultaneously to both sides, for label application, e.g. to place labels at an inclination of 45 degrees. By employing a longer plunger arm it is possible to apply labels into deep recesses, or even into the bottom of something like a cup or glass.

Air jet/air blow is another frequently used method of label application in which the labels are stripped as usual and then, after release, are again held on a low vacuum, but this time are blown onto the container by applying air pressure.

Using this method of label application it is possible to apply labels in positions that would be virtually impossible by other methods. Absolutely no damage will be done to the surface or product to be labeled.

A variation of air jet application is where the application head is built into a mechanical arm which can then be positioned to label by air jet, products such as television tubes.

LABEL AND PRODUCT CONTROL
To ensure that labels are applied accurately and consistently to products, the self-adhesive label applicator may be fitted with various label or product tracing/control devices.

Label tracing devices monitor the gap between the

individual die-cut labels and interrupt the web-feed at the end of each label. This resets the system to receive the next start signal from the product tracing devices, which include micro-switches, photocells or spot color readers. Other controls on the web may be used to detect missing labels, the end of the reel or web breaks.

With many applicator systems the actual product or container triggers the application of the label through the use of product tracing/ control devices. This means that even with erratic or irregular product flow the system will automatically compensate and only dispense a label as required. Many different scanning/sensing technologies are available on self-adhesive applicators, from photocells with reflector mirrors, transmitter/receivers, proximity switches, spot or color readers, or for certain applications, micro-switches. For semi-automatic applications, foot-switches may be used.

ADVANTAGES AND DISADVANTAGES OF SELF-ADHESIVE LABEL SYSTEMS

The advantages and disadvantages of self-adhesive labels as a decorative label technology are summarized here;

Advantages

- Clear self-adhesive filmic substrates can achieve a 'no label' look on clear containers
- Receptive to high quality graphics/ embellishments (combination printing)
- Wide selection of substrates and adhesives
- Compatible with plastic containers for recycling
- Quick change-overs facilitate logistic flexibility and inventory control and therefore suited to short run, just in time applications
- Application efficiency – quick changeovers, accuracy
- Fast conversion and finishing time
- The diversity of applications is impressive. It includes underwater labeling, high and low-temperature labeling, also labels for sterilizing and autoclaving.

Disadvantages

- High cost and environmental issues related to the disposal of backing liner material
- High substrate cost
- Slower application than wet-glue on some jobs.

For example the application rates may not be fast enough to deal with today's modern filling lines for soft drinks and beers, although this is becoming less of an issue.

LINERLESS LABELING

As already discussed conventional pressure-sensitive labels consist of a face material, an adhesive and a siliconized release liner backing that functions as a carrier for the label and protects the adhesive layer during handling, printing, finishing and right up to the point that the label is ready to be applied.

Applicators and labelers for conventional pressure-sensitive labels need to peel away the backing release liner for each label to be dispensed and successfully apply it to a container, product or pack. The liner then needs to be disposed of as waste.

Conventional pressure-sensitive laminated labels like this produce the highest waste levels of any packaging component, with over 50 percent lost during conversion and final end-use application.

Linerless pressure-sensitive labels however enable the release liner to be eliminated.

Indeed, the development of the linerless process was driven by the need to remove the backing liner used in pressure-sensitive labeling, in order to minimise wastage.

Linerless pressure-sensitive label technology is not new, it has been around for over 35 years. Today the improved techniques being used and developed have coincided with a greater awareness of the importance of sustainability and the need for retail chains and brand owners to prioritize the environmental issues facing the packaging industry. Millions of tons of used silicone-coated paper and filmic release liners are land filled each year and this is not good environmental practice.

Today, linerless labels are most commonly found in the form of pressure-sensitive labels for the blank label industry and as thermal labels used in print and apply weigh-price label dispensers. They are also popular in market sectors such as food and logistics, but the impact and growth in the other label markets has been slow. That situation is now beginning to change as more countries start to tax liner waste as packaging material rather than industrial process waste and it is becoming more difficult and complicated to dispose of the liner. Some of the major packaging users have the ambitious objective of ensuring 100 percent of its packaging designs are reusable, recyclable or suitable for good environmental waste management.

Linerless technology ticks many of the sustainability packaging requirements and with improved coating technology and advanced applicator equipment, there is the potential to impact on several prime label sectors, particularly in food packaging and the labeling of glass bottles and jars.

Manufacturers of packaging converting equipment are now realising the full potential of the linerless decoration system.

HISTORICAL CONTEXT - MONOWEB

Monoweb introduced in the mid 1980's by Waddington PLC was the first truly linerless pressure-sensitive label.

A silicone coating applied over the surface of the printed substrate allowed a non-contaminated release from the reel. The reel of printed substrate was positioned into the applicator and the profile shape of the label being applied was die-cut at the point of application and then removed from the web by direct application to the container.

There were some drawbacks with the Monoweb system. Any problems with the die-cutting operation immediately stopped the line and the quality and durability of the die tooling was critical. The system also introduced a new set of skills for the line operators, which required additional knowledge and operating skills, particularly with the die-cutting process, in order to extract the best performance out of the cutting tool.

The system of print to die-cut registration was achieved using a sprocket punched hole system using pins which located into the punched holes to ensure the correct registration between the printed image and the die-cutter. The punched holes located at the edges of the web required removal using slitting and waste extraction.

THE EVOLUTION OF THE LINERLESS LABEL

The development of the linerless labeling has evolved considerably in recent years.

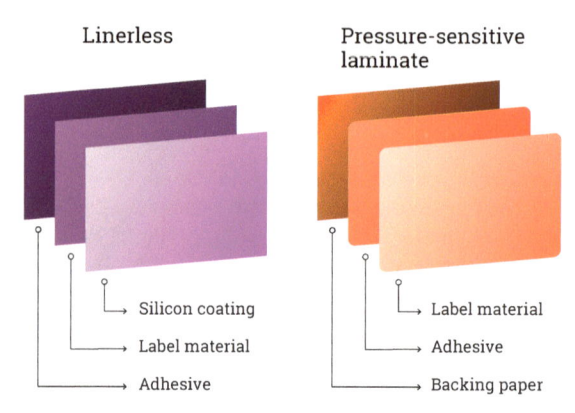

Figure 4.21 The structure of a linerless label versus that of conventional pressure sensitive laminate

Linerless labels use the same adhesive method as a pressure sensitive label, in that the label has an adhesive layer on the back of the label substrate. There are however some fundamental differences.

A conventional pressure-sensitive label has a face substrate positioned onto a siliconized liner ensuring that the adhesive and face stock do not stick to each other in the roll. The liner also acts as the carrier for the printed and profile cut labels which are dispensed onto the product direct from the liner.

In the linerless process the face stock is printed as a single layer and the label surface is coated with a silicone layer and the adhesive coating applied to the back of the label stock.

Alternatively a separate liner is coated with the adhesive and laminated onto the printed label stock allowing the adhesive to transfer from the liner to the back of the label stock.

The liner and face stock are then de-laminated, the liner is re-wound for reuse leaving a single layer of printed face stock, which is then rewound into a reel. Figure 4.21 compares the structure of a linerless label with that of conventional pressure-sensitive laminate.

The silicone coating on the face on the printed label, prevents the adhesive from adhering to the face of the label (Figure 4.22).

The removal of the need for the backing liner offers

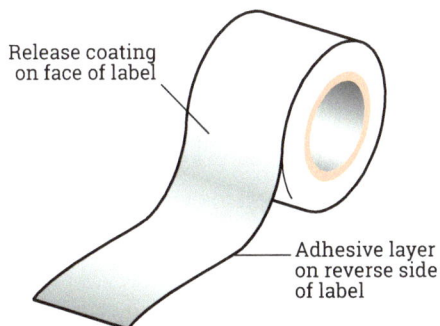

Figure 4.22 Linerless labels can be regarded rather like a large roll of adhesive tape

Figure 4.23 ILTI linerless applicator (left) on a production line

significant environmental and cost benefits.

Release liners used in conventional self-adhesive systems generally are not recycled and the bulk of them find their way into landfill. By removing the need for a release liner the amount of waste material is significantly reduced and therefore the costs related to waste disposal, storage, transportation and substrate usage are drastically cut.

The removal of the release liner also increases the number of labels per roll resulting in fewer roll changes for press operators and also a reduction in downtime on the filling and label application lines. The increased number labels per reel also reduces the amount of inventory storage space required for the printed label stock.

PRINTING AND CONVERTING LINERLESS LABELS

A common misconception with the linerless system is that special printing equipment is needed to print them. This is not the case and labels are printed using conventional printing machinery.

A typical linerless press configuration will comprise of a single unwind unit, flexographic printing units and a rewind unit. As there is no liner, printing on a single side or both sides of the label substrate can be carried out.

When printing is completed the reels are ready for the second stage which is the application of the adhesive and the silicone coating which also helps to

protect against UV and moisture exposure.

Next is the die-cutting process, waste removal and finally the checking/slitting operations.

Individual coating units are used for the hot-melt adhesive and the silicone coating process.

A base liner substrate is coated with the adhesive, at the same time the UV silicone coating is applied to the printed side of the face-stock.

The adhesive coated liner is then laminated to the reverse side of the printed face stock allowing the adhesive coating to be transferred to it. The laminated web then proceeds into the die-cutting unit, the waste matrix is removed and the liner and face stock are separated.

The liner is rewound ready for re-use and the finished linerless printed reels are slit and wound into smaller reels ready for application on the filling or labeling lines.

This method of linerless label manufacturing makes the labels suitable for the following end-use applications.

- Print and apply labeling which require variable data etc. as used in product distribution and logistics tracking
- Primary product decoration of a product or container which does not require a profile shape i.e. best suited for applications requiring square or rectangular cut labels.

APPLYING LINERLESS LABELS

It is at the application stage where additional equipment is required. A special cutting unit is required which separates the label ready for application.

To apply the label a mechanical or laser cutting system cuts/slits a single label from the adhesive coated linerless reel.

After the cutting process has been completed the label is applied by adhering the leading edge of the label to the container and then wiping the remainder of the label onto the container.

For difficult and uneven surface applications, the single label can be blown onto the container surface using suction to hold the label and air jet to apply the label.

Figure 4.23 shows a linerless applicator positioned on the left side of the filling line cutting and applying pre-printed linerless labels.

One of the big differences between linerless and pressure-sensitive labels is that linerless labels cannot be profile die-cut during the printing operations.

As explained previously the printing and embellishing processes are carried out using only the single layer of face paper and not the sandwich of substrates used in the conventional pressure sensitive label system.

Because the labels are printed in a continuous reel with no liner to support the individual profile cut label, this means that there are limits on the profile shape of the linerless label.

This limits the label shape to a rectangular or square cut profile. Developments however are taking place to overcome this issue of profile shapes produced in a linerless format.

Because of the limits on the shapes which can be used, linerless systems often use clear film to create the effect of a profile shaped label. The design of the printed image is used to create the effect of a profile shaped label albeit that the label may be square or rectangular in shape.

ADVANTAGES AND DISADVANTAGES OF LINERLESS SYSTEMS

A summary of the advantages and disadvantages of linerless systems is summarized below;

Linerless Advantages
- No release liner required
- Printing, adhesive application, die-cutting and rewinding of finished reels takes place in one pass
- Eliminates the problem of waste removal and re-cycling of the liner material
- Minimal waste matrix required making a material cost saving on face substrate
- More labels per roll and less roll changes during application
- Lower reel weights and transportation costs

Linerless Disadvantages
- Investment is required for the linerless application equipment
- Some materials are unsuitable for linerless application
- The printer has to apply silicones and adhesives (hotmelt)
- The printer will require suitable coating units
- Filler/labeler will require suitable applicator to apply the linerless labels.

THE CATCHPOINT SYSTEM – A CASE STUDY

A number of companies offer proprietary linerless technology and applicator systems for both primary and secondary product decoration.

These companies include Catchpoint, Ravenwood Packaging, ETI Converting and Catchpoint licensee ILTI srl, a review of linerless technology by manufacturer is contained in the Label Academy book on Label Dispensing and Application Technology.

One of the systems which has been developed and patented is the Catchpoint system, this technology has been designed specifically for the application of linerless labels.

The system uses the same linerless manufacturing process explained earlier, but incorporates calibrated micro perforations (0.2 or 0.3 mm ties) positioned between each label. The perforations secure the labels within the web/reel but allow the label to be individually separated from the printed reel and accurately applied to the container at high speed, these perforations are known as Catchpoints™.

Catchpoint have worked closely with a number of

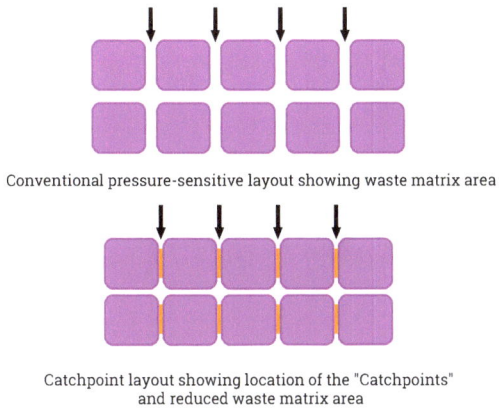

Conventional pressure-sensitive layout showing waste matrix area

Catchpoint layout showing location of the "Catchpoints" and reduced waste matrix area

Figure 4.24 Comparison of conventional pressure-sensitive layout with the Catchpoint layout

material suppliers, printers and manufacturers of labeling applicators and are one of the first companies to develop a practical linerless label application system with none of the efficiency and safety issues experienced with the earlier Monoweb system.

Labels are shaped by a combination of die-cutting and micro perforations which are incorporated into the die-cutting tool and give an accurately repeated label shape. There are however limits on the shape of the label that can achieved using this method of cutting but shaped cutting of the top and bottom of the label can be easily carried out whilst still maintaining the perforated Catchpoints which are located in the non-profile cut areas.

Because the labels are linked together by the Catchpoints, spacing between the labels is not necessary and therefore eliminates the waste matrix which is normally associated with conventional self-adhesive labeling. Figure 4.24 below compares a conventional pressure-sensitive label layout to space saving layout required by the Catchpoint system.

CATCHPOINT LABEL APPLICATION
The Catchpoint application system is a simple way of applying labels to the container. The labels are held together in the web by micro perforations and at the

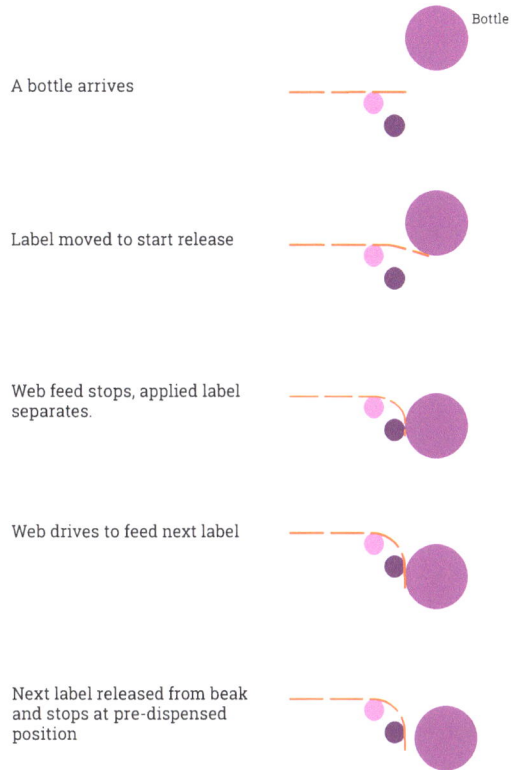

A bottle arrives

Label moved to start release

Web feed stops, applied label separates.

Web drives to feed next label

Next label released from beak and stops at pre-dispensed position

Figure 4.25 The Catchpoint label release sequence

point of application the labels are separated by momentarily pausing the web over a stationary separator edge (or in the ILTI version a moving separator tool) so that the label being applied is separated at the Catchpoint. This separates the label from the stationary web thereby allowing the single label to be accurately applied. Figure 4.25 illustrates the Catchpoint label release sequence.

The peel plate or beak used for conventional pressure-sensitive labeling are replaced by a label separation system, labels can be separated and applied at high speed.

LINERLESS FROM A LAMINATE
Filmic substrates are a rapidly growing part of the pressure sensitive label market, Catchpoint has worked

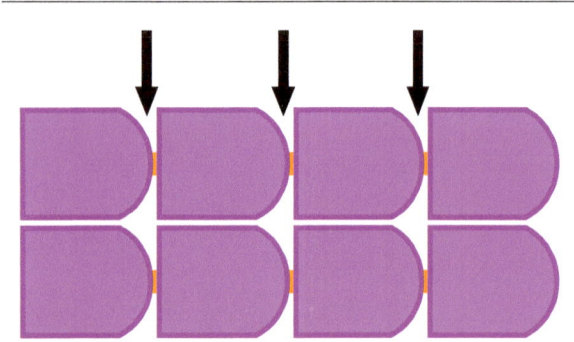

Figure 4.26 Catchpoint labels showing the position of the perforations for a LeanLean profile cut label application

with laminate suppliers in the development of thin filmic labels which can be used in the linerless system.

The filmic face substrate is printed conventionally, but after printing the clear film base liner is de-laminated from the back of the printed face substrate and transferred onto the top of the face substrate. This leaves the adhesive layer on the back of the face substrate thereby creating a linerless label constructed with two layers of very thin filmic substrates and leaving the original release coating, which is on the liner to provide the release coating needed for a linerless format. The major advantage of this system is that it allows the use of filmic substrates with a much reduced face film calliper (down to 20-30 micron).

Linerless labels do not required the same stiffness characteristics that are required by conventional pressure-sensitive labeling. The Catchpoint application allows the label being applied to be attached to the container whilst the printed reel simultaneously gives stability and control of the thin label at the point of application.

CATCHPOINT LEANLEAN™
Catchpoint has extended the use of the very thin films to both enhance sustainability and deliver label shapes which cannot be provided in a linerless format. Catchpoint LeanLeanTM is a conventional PS label delivered from a filmic liner. The new system overcomes the stiffness limits required by the conventional

application by the use of limited catchpoints between the label images as shown in Figure 4.26. The minimal ties release the follower label to a defined pre-dispense and this label is then controlled by the next arriving container to repeat the sequence.

This system uses a standard pressure sensitive filmic substrate construction which is printed and embellished using conventional methods, the die-cutting process which includes the catchpoints is the next process during which time the label is profile cut and the reduced waste matrix for the shape only is removed.

SUMMARY OF CATCHPOINT ADVANTAGES AND DISADVANTAGES
Catchpoint - Advantages
- No liner or carrier is required in a linerless format
- A good range of application options across the whole speed range
- The ability to offer very thin lower cost face materials as a conventional label with reduced matrix waste
- Applicator change unit suitable for use on existing application lines
- A label application system using micro-perforations which gives accurately applied label placing
- System uses existing conventional die-cutting systems
- A range of profile shapes which meet many existing label shape specifications on both paper and clear films
- Fewer web breaks and a claimed 2-3 percent improved line efficiency.

Catchpoint - Disadvantages
- The printer cannot apply silicone and adhesive as efficiently as the major laminators
- Linerless application systems can be more complex than self-adhesive systems
- Some very special materials cannot be used
- New investment is required to convert linerless labels and to install special application systems
- The cost of linerless labels may be higher

Chapter 5

Glue applied labels

For decades, the most popular and widely-used method of decoration for glass and cans has been the use of wet glue paper labels. These labels are predominantly printed using the lithographic and gravure printing processes and increasingly the UV flexographic printing process. The labels are usually applied on the filling line using a wet-glue or hot melt adhesive.

Paper manufactures now provide label paper specifically developed for the beers, wines, spirits (Figure 5.1), water and soft drinks markets and these can be applied in the traditional cut and stack 'one by one' format or on reel to reel systems.

The key uses for glue-applied labels are in the high speed, high-volume, low changeover labeling of drinks bottles and for canned foods – both human and pet food – where application speeds up to 60,000 – 80,000 or more containers per hour are available.

Although used mainly for primary decoration some wet glue labels are also applied as top straps over the closure as a tamper evident or security feature.

Figure 5.1 Wet-glue labeling is a cost effective way of applying highly decorative labels to glass bottles

WET-GLUE LABELING SYSTEMS

The wet-glue labeling system comes in three main formats:

- Punched labels where the labels are ram-punched into specific profile shapes, which are compatible with the container profile. These are also referred to as 'patch labels'.
- Cut labels which are rectangular or square in shape. This profile is suitable for 360 degree wrap- round decoration or as patch labels.
- Roll-fed labels, supplied in reels for 360 degree decoration. These are unconverted labels and are not cut to shape, but cut singly from a reel at the point of application.

The majority of wet-glue applied labels are manufactured by a conventional multi-stage operation and are printed and embellished in sheet form (Figure 5.2). In the case of high-volume gravure printed wet-glue labels the printing would be carried out on a web fed gravure press and the printed reels would be transferred onto a sheeting unit which can be an in-line or off-line. There is however an increasing use of medium size (350-650mm web width) multi-process web-fed presses which print and embellish in one pass.

The first stage of this operation is the production of sheets of printed labels. If foil blocking, embossing, bronzing, or over-varnishing is required then the printed sheets are further processed through each

Figure 5.2 Heidelberg Speedmaster multi-stage sheet-fed printing press

Figure 5.3 Blumer Strip cutting and ram punching unit profile cutting a wet-glue label

separate stage. The finished sheets of labels are then guillotined, and square cut or profile punched (Figure 5.3), batched, packaged and palletized.

MULTI-PASS SHEET FED WET-GLUE LABEL MANUFACTURING VERSUS SINGLE PASS MANUFACTURING

Manufacturing using multi-pass processing can have a production cycle of several days, or in some instances several weeks dependant on the quantities required, with perhaps 10 people involved in the processes as sheets pass through the printing, embellishing and converting processes.

The length of this extended manufacturing window is a problem, with brand owners demanding more frequent short run orders and quicker delivery schedules. The problem of this lengthy manufacturing window has led to the use of web-fed combination/platform presses.

These presses have evolved via the self-adhesive industry and can replicate the printing, embellishment and finishing techniques used for conventional wet-glue label manufacturing. Multiple processes can be combined using all the major printing and embellishing processes i.e. offset litho, rotary screen, UV flexo, gravure and digital printing plus foiling, embossing, laminating and converting into single

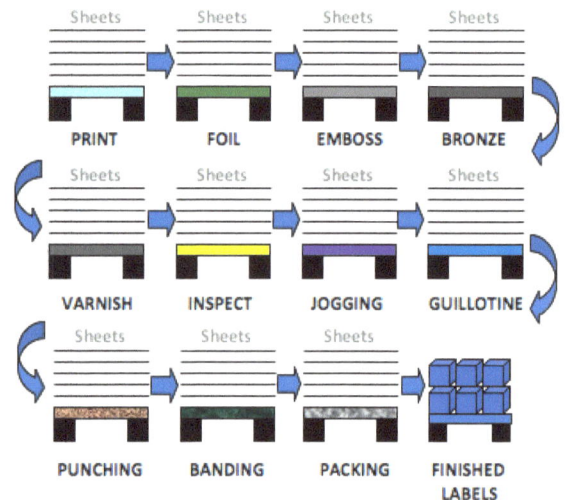

Figure 5.4 Standard sheet fed wet-glue label manufacturing with multiple passes

labels, all in one pass.

Figures 5.4 and 5.5 illustrate the effect that one-pass printing and converting has on the manufacturing window. The significant difference is the removal of multiple passes when labels are printed in sheet-fed format.

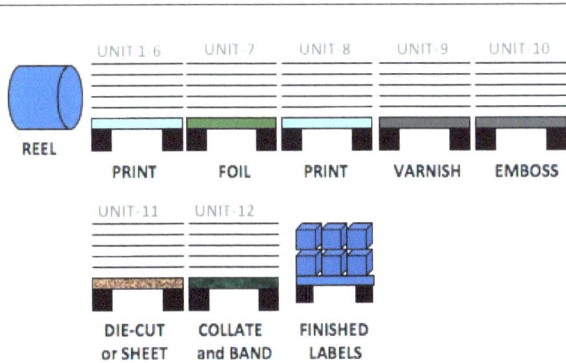

Figure 5.5 The in-line reel fed one pass wet-glue label manufacturing process

SUBSTRATES USED FOR WET-GLUE APPLICATION

Paper face-stocks such as Chromolux 60, 80, 90gsm materials are commonly used but increasingly popular forms of wet-glue paper labels are foil-laminates and more recently metalized papers. These offer a reflective metallic finish, which can be printed on, offering excellent design opportunities. Metallized paper, as opposed to foil-laminates, uses less aluminium and could be said to be more environmentally friendly, using less non-renewable natural resources.

The label papers are available in standard, wet strength, embossed and high gloss versions to cover all the technical and marketing requirements and are suited to a wide variety of other applications, such as box covering, lamination, food wrapping and interleaving.

TYPICAL WET GLUE PAPER SUBSTRATES

The main types of wet-glue paper materials and their characteristics are summarized here;

Uncoated Paper
- Litho
- Vellum
- Uncoated surface

Uncoated characteristics
- Low gloss level
- Greater absorbency
- Less suitable for quality image reproduction

Coated Paper
- Matt coat
- Semi-gloss
- Gloss
- Cast coated

Coated characteristics
- Coated surface
- Gloss/reflective surface
- Lower absorbency
- More suitable for quality image reproduction

When selecting a substrate for use in wet-glue label applications the following factors must be considered;
- Direction of the grain and the resulting effect on label curl in uncontrolled humidity
- Substrate moisture content during storage and transportation
- Humidity levels in the printing environment
- Humidity levels in the filling and label application areas

DECORATIVE FINISHES

High gloss lacquers are used on the face of wet-glue applied labels for scuff resistance and increased shelf impact. For lithographic printing, the norm is the use of ultra violet (UV) lacquers, whilst for photogravure printed labels, a range of solvent and water-based lacquers are used. A combination of matt and gloss lacquers can give increased shelf impact.

Embossing is used where a specific part of the design is raised to enhance the appearance of the label.

With photogravure printing, in-line all over or partial embossing is possible, producing a tactile feel to the label in addition to helping improve line efficiencies, especially in relation to metalized paper labels.

Foil-blocking is used where parts of the design can appear in a high gloss reflective foil. This is available in a wide range of colors.

WET-GLUE LABEL APPLICATION

In general terms this system of label application

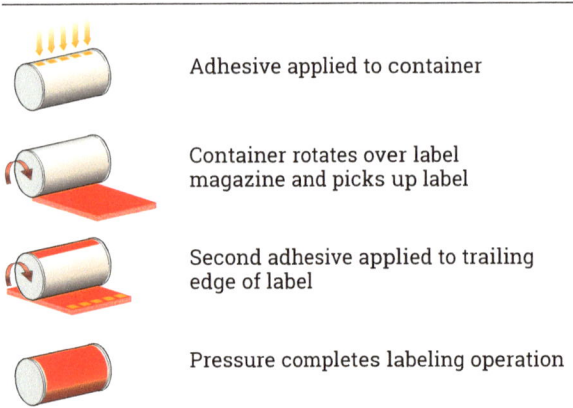

Adhesive applied to container

Container rotates over label magazine and picks up label

Second adhesive applied to trailing edge of label

Pressure completes labeling operation

Figure 5.6 Stages involved in labeling a can

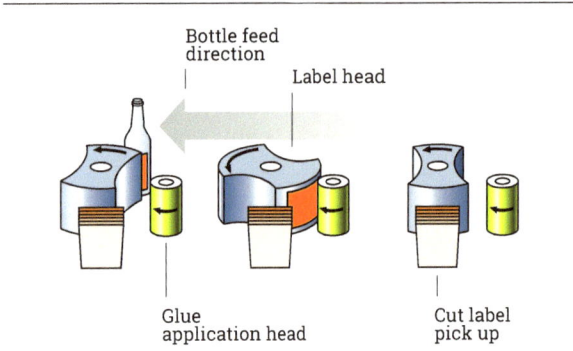

Bottle feed direction

Label head

Glue application head

Cut label pick up

Figure 5.7 A typical patch label application sequence

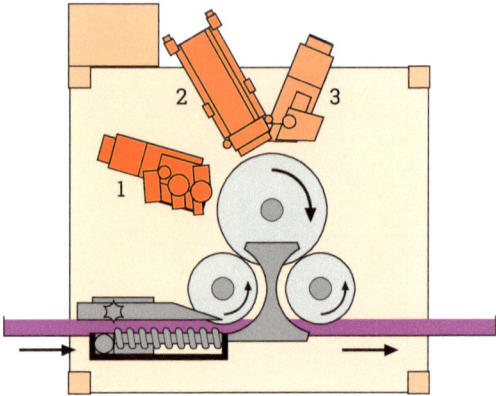

Figure 5.8 Single label leading and trailing-edge gluing using a hot melt double head
Key: 1 Hot melt gluing head, 2 Pre-cut single labels in magazine, 3 Hot melt gluing head

The stages involved in the labeling of a can are illustrated in Figure 5.6.

The application of a typical patch label is illustrated in Figure 5.7. In simple terms the cut label is picked from a stack, held in place by a vacuum, whilst wet-glue is applied to its reverse by a glue application head and before it is transferred to the bottle or container.

Figure 5.8 shows a system operating two hot melt gluing units. Gluing unit 1 applies a vertical strip of glue to the container. The container revolves as it passes the label magazine and the label is then picked from the magazine by the glued strip. Gluing unit 2 applies glue to the trailing label edge and the label is wrapped around the container. The label is glued together at the overlap.

Figure 5.9 shows the positioning of strips of adhesive on the leading and trailing-edges of the label. This method uses much less adhesive than systems that apply adhesive all over the label.

WET-GLUE WRAP AROUND LABELING
The format of the wrap around label is exactly as the wording describes i.e. the label wraps fully around the

places a single cut label from a stack of labels, direct into the container using either wet-glue adhesive or a hot melt adhesive system.

The adhesive can be applied to either the full area of the reverse side of the label or by gluing the trailing and leading edges of the label. The type of adhesive application can also be selected from options such as skip, pattern or stripe, depending on adhesion, application speed or drying speed.

The key methods used for applying a wet-glue label are simply explained here.

In can labeling adhesive is first applied as a stripe to the container and a second stripe of adhesive is applied to the trailing edge of the label as it rotates over the label magazine.

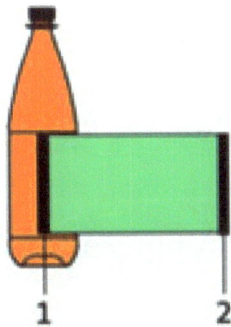

Figure 5.9 Wrap round label featuring hot melt glue strips on leading and trailing edges

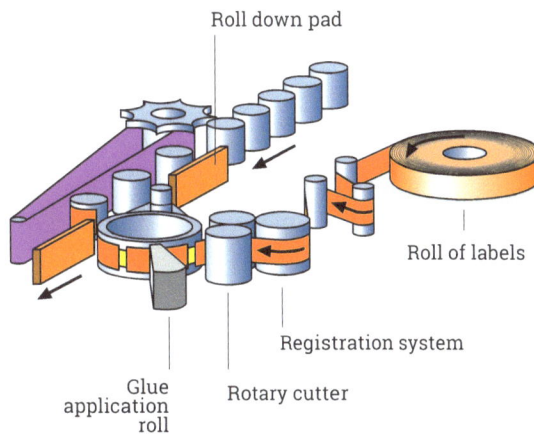

Figure 5.10 Typical reel fed wrap round applicator

container giving 360 degree graphics. Labels are usually rectangular in shape and are very popular for decorating bottles, jars and cans.

Application can be from a stack or a reel and usually features an overlap which allows the leading edge and the trailing edge to be glued together.

A key advantage of the wrap around label is they are cost effective as the container can be decorated with a single label and not multiples as some applications are i.e. body, neck and back labels.

Figure 5.10 illustrates a typical reel-fed wrap round label application.

WET-GLUE – ADVANTAGES AND DISADVANTAGES

Wet-glue paper labeling offer the following benefits:
- The lowest label cost of all methods decoration
- High-speed application rates, exceeding 1,000 bottles per minute
- Excellent ability to reproduce photographic designs, tone-work, metallic colors, etc.
- Relatively low design origination costs for lithographic plates
- No waste matrix or liner to dispose of
- Easy removal of the labels from returnable bottles in a caustic wash bath.

The constraints of wet-glue labels are as follows:
- Multi-stage manufacturing with longer lead times
- High cost of application equipment – including change parts and adhesive
- Cannot be applied to highly-shaped areas of containers
- Paper labels have limited moisture resistance compared to plastics labels
- Poor product and damage resistance (paper).

SUMMARY

Wet-glue labeling continues to dominate in the high volume sectors for bottle labeling, notably beer and spirits, mineral water and canned foods. More self-adhesive suppliers are however, competing in the traditional wet-glue markets, particularly in the wine and spirit sectors and in those markets that use filmic substrates.

Many end users that have invested in wet-glue application systems are reluctant to change to self-adhesive for high volume run lengths.

Wet-glue labeling in comparison to the self–adhesive labeling system has much higher application speeds facilitated by label substrates specifically developed for high speed application.

Chapter 6

In-mold labels

In-mold labeling (IML) is a decorating technique used worldwide for blow molded bottles as well as injection molded and thermoformed containers, or other plastic objects.

In-mold labeling is a process that allows for the decoration of plastic containers as part of the molding process. In simple terms a pre-printed paper, synthetic or film label which is placed inside the mold before the plastic is blown, injected or thermo-formed to produce a plastic bottle or tub.

The hot plastic and the label merge making it an integral part of the molded container without any label edge visible.

It was pioneered in Europe for injection molding in the early 1970's and in the United States for blow molding later in that decade.

What differentiates in-mold labels from conventional glue-on labels is the heat seal coating that is applied to the back side of the in-mold label stock during the manufacturing process. Labels for injection molded purposes however do not require an adhesive on the back side.

In Europe, some 80 percent of in-mold labeling is with injection molded or thermoformed tubs for dairy foods, such as soft spreads, margarines, cheeses, sauces and ice cream, with 20 percent used with blow-molded containers for under-the-sink products, household chemicals and industrial, laundry products, personal and hair-care products and some juices (Figure 6.1).

In the States, it is in-mold labeling of blow-molded containers which dominates the market.

The in-mold labeling process has found increasing application, meeting the demand forprecision labeling of high-quality, low weight molded containers using the same type of material as that of the container (e.g.

Figure 6.1 Examples of in-mold labeled products

PP on PP) for ease of recycling. The process is economical only if extremely large production quantities are involved.

Because of the high cost of the basic molding equipment and molds, plus the need to modify these to be able to insert and position labels accurately into the mold, in-mold labeling has had a somewhat limited acceptance in the market place and IML has little more than a 2-3 percent market share.

PRINTING AND CONVERTING

The process used for the manufacture of in-mold labels is identical to the process used for wet-glue labels production.

The majority of IML labels are produced in a sheet-fed format, but some are also produced in the reel. The printing is predominantly done using the litho, gravure or more recently the flexographic process

If required the sheets can then be foiled blocked and over varnished. The printed sheets are then converted by guillotining into strips or gangs and square cutting or profile ram punched using the same method as that used in wet-glue labeling.

The labels are then banded, packaged and palletized ready for the container molding line. It is now common practice to also convert the label as an in-line operation using a rotary die-cutting and label collection system either linked to the press or as an off-line operation.

The digital printing process is also being used for in-mold label manufacture. The variable repeat length of digital printing allows the production of very large labels which are used for the bigger containers and tubs. Digital printing is also a valuable tool when used for the purpose of rapid prototyping, when new product containers are at the development stage.

Inks are critical to the process with labels protected on the surface with a UV or EB curable top coat.

Die-cutting is also critical, particularly when labels printed in sheets are stacked and rammed through a tunnel, emerging cut to size. Edge welding may be an unwanted consequence of this process.

MOLDING AND DECORATION OF THE CONTAINER

Molded containers are usually made of high density polyethylene (HDPE) and polyester (PET). Open top containers made of PP are also injection molded and labeled by IML. Many plastic articles, including toys and automotive parts are also in-mold decorated to add value or provide warning or instructional labeling.

THE IML PROCESS

There are three types of molding system used for in-mold decoration, each one differing in the way the molding process is carried out.

BLOW MOLDING

During the in-mold labeling process, a label or appliqué as it is sometimes referred to, is placed in the open mold and held in place against the internal wall of the mold by vacuum ports, electrostatic attraction, or other appropriate means. The mold closes and molten plastic resin is extruded into the mold where it conforms to the shape of the object. The completed container is then released from the mold.

Figure 6.2 illustrates the blow molding IML process

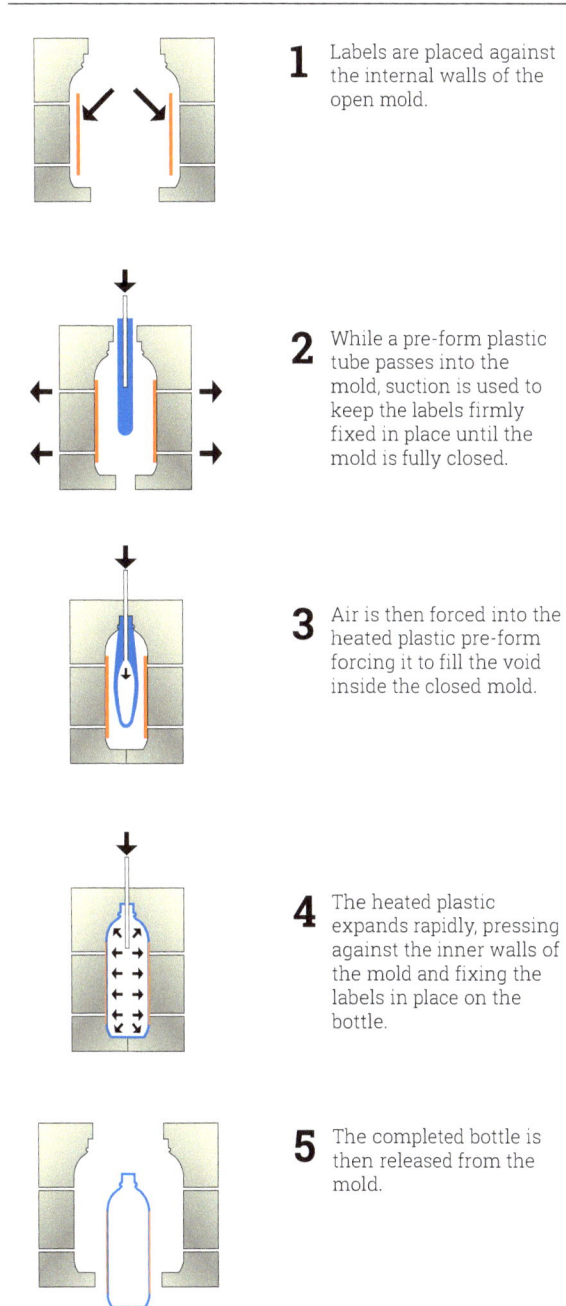

1 Labels are placed against the internal walls of the open mold.

2 While a pre-form plastic tube passes into the mold, suction is used to keep the labels firmly fixed in place until the mold is fully closed.

3 Air is then forced into the heated plastic pre-form forcing it to fill the void inside the closed mold.

4 The heated plastic expands rapidly, pressing against the inner walls of the mold and fixing the labels in place on the bottle.

5 The completed bottle is then released from the mold.

Figure 6.2 The operations involved in in-mold labeling

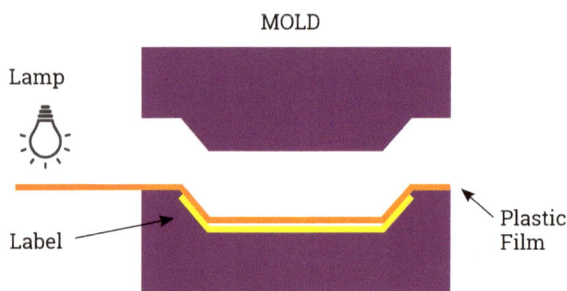

Figure 6.3 In-mold labeling using the thermoforming process

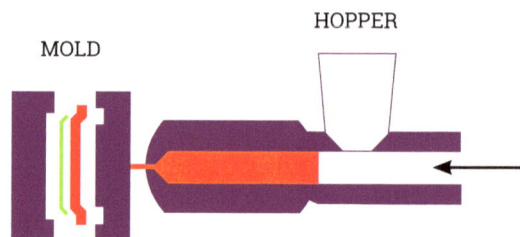

Figure 6.4 In-mold labeling using the injection molding process

graphically.

THERMOFORMING

Thermoforming differs in that the label is placed in the mold and a sheet of plastic is then placed into the mold, positioned on top of the label. The heated mold then closes and forms the container shape with the label integrated into the container. This system is widely used for the manufacture and decoration of container lids (Figure 6.3).

INJECTION MOLDING

Injection molding works as the name describes. The label is placed into the mold and the heated plastic is injected into the mold thereby forming the container shape.

Whilst the molding of the label into the liquid plastic is taking place plastic granules are fed into a screw tool that takes the granules forward towards the mold.

The screw assembly is heated to melt the plastic ready for injecting into the mold. After injection the plastic quickly cools and the decorated container is ejected and the process repeats (Figure 6.4).

SUBSTRATES

Initially, the IML process involved paper labels coated with a heat-seal back layer which fused to the container during the blow molding process. Whilst some paper labels are still used, the market is moving quickly towards filmic label materials that are compatible with the primary container, thus allowing total pack recyclability.

More recently, synthetic paper materials such as Polyart or Synteape have become more common, as well as specially developed OPP films which fuse directly to the container and eliminate any tendency to an orange-peel effect. These substrates are used for the labeling of injection molded containers. All the in-mold label substrates must have good lay-flat properties for trouble-free feeding from the molding machines.

The control of surface-printable IML filmic substrates throughout the production process is difficult and requires good quality process control to overcome curl and scratch issues.

Static electricity can be a difficult problem during the printing, converting and application operations because the films behave differently than standard IML films.

The key substrate properties for in-mold labels can be summarized as follows;
- Generally filmic/synthetic substrates
- Range of substrate thickness 50-1000 micron
- Range of substrate finishes matt or gloss, high gloss, gold and silver metallics
- Transparent films used for "no-label" look
- No slip surface and good lay flat properties

- UV or EB top coat for protection

ADVANTAGES AND DISADVANTAGES

One of the issues that arise with IML is the lack of flexibility to quickly change the container decoration.

As the finished container is pre-labeled it becomes important that unit quantities are accurate. Any rapid change of decoration to a specific brand is difficult without incurring obsolete containers.

This situation also applies to the length of time required to produce the IML labels. The printing and finishing of the IML label is exactly the same as wet-glue label manufacturing but the process of molding and decorating of IML containers is much slower and will therefore extend the manufacturing window of IML products.

The key benefits of IML are summarized below;

- IML offers label and image durability in harsh environments.
- The label is an integral part of container thereby increasing the container side wall strength. This in turn facilitates the use of lighter weight containers resulting in 10-15 percent saving in plastic.
- Resistant to aggressive and difficult products e.g. chemicals/fats.
- Filmic labels are often compatible with container thereby aiding recyclability.
- Simplifies the filling and packaging operations by eliminating a major source of efficiency reductions i.e. the labeling process which represents the major cause of filling line stops.
- IML transfers the problem of applying the label from the filler to the container manufacturer.
- Pre-labeled containers result in increased packaging line speeds with opportunities for in-case filling.

The key constraints of IML are summarized below;

- The IML process is only suited to plastic container manufacture
- IML has to be done as part of molding process adding to cost of container production and increased cycle times
- The high capital cost of the equipment and the molds can be prohibitive
- There can be long lead times for labels and molds and high wastage factor during molding
- Potential to create obsolete stock of pre-labeled containers
- The disadvantages of IML tend to shift to the container manufacturer and blow molder who will incur longer manufacturing cycle times.
- The labels themselves are more expensive than wet glue applied labels, but this on-cost is typically offset against higher efficiencies on the filling line.

IN-MOLD LABELING (FORM FILL SEAL) (ERCA)

In yoghurt packaging, which is dominated by the use of styrene, the successful introduction of the ERCA form-fill-seal machine has completely revolutionized container decoration. Rather than being directly printed by the container manufacture, a large proportion of yogurt pots are now packed in clusters and are in-mold labeling in aseptic conditions at the point of filling, using a heat sensitive pre-printed wax coated paper which adheres to the pots during the thermoforming process.

In this instance in-mold labeling allows light weighting of the container walls.

Chapter 7

Sleeving

Sleeve labels are typically produced from polymer film substrates capable of shrinking biaxially around the product when subjected to heat.

Figure 7.1 Typical shrink sleeve application

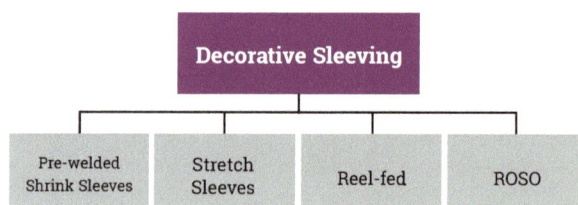

Figure 7.2 Summary of the key decorative sleeving formats

They provide 360 degree graphics around a pack offering surface protection from rubbing and a high degree of label security. Indeed, the process has proved ideal for additionally shrinking the film around the cap of a bottle or jar so as to provide evidence of tampering.

Sleeving as a primary decoration method has achieved significant penetration into a number of market sectors, notably beverages, household, foods, pharmaceuticals, toiletries and cosmetics markets (Figure 7.1), and currently is the fastest growing of all the labeling processes.

In the beverage sector it is used as a light-weighting vehicle for glass manufacture. On plastic bottles the advantages of film based sleeves that are compatible with the primary container have been recognized as an aid to re-cyclability.

In recent times innovative shrink films, new printing techniques and printing finishes and the development of efficient application machinery and shrink tunnels, have all contributed to the continued growth in sleeve decoration.

This chapter is designed to help the reader understand the different shrink sleeving processes and methods whilst maximizing the benefits derived from this exciting product decoration method (Figure 7.2).

There are 4 main sleeving formats used today:
- Pre-welded shrink sleeves
- Stretch sleeving
- Reel-fed wrapround sleeving
- Roll-on shrink-on sleeving (ROSO).

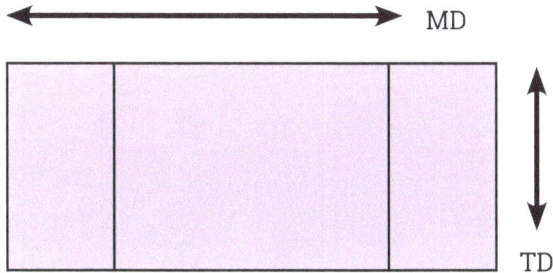

Figure 7.3 Direction of shrink machine direction (MD) v transverse direction (TD)

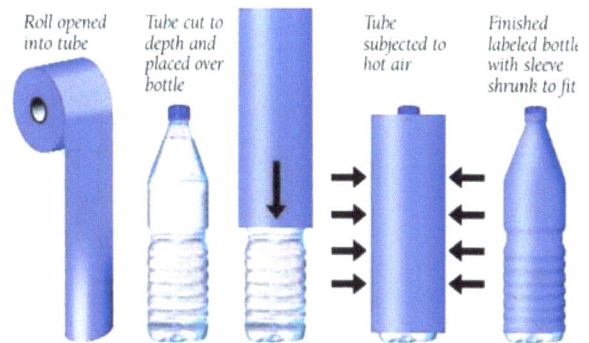

Figure 7.4 Principle of shrink sleeve labeling

Each of these decoration methods will be dealt with in this Chapter.

SHRINK SLEEVING – SUMMARY OF BENEFITS
The benefits of shrink sleeving include the following factors;
- 360 degree full length decoration
- Translucent colors, or transparent windows to view contents
- Total container coverage protecting against contamination
- Inventory reduction – containers can be labeled on-demand
- Tamper evidence
- Matt or gloss finishes, or a combination of both
- Metalized, pearlized, photochromic and thermo-chromic finishes
- Tactile sleeves
- Ideal for labeling complex shaped containers
- Ability to incorporate promotional ideas
- An effective alternative to printed/coated containers.

PRE-WELDED SHRINK SLEEVES
Simply put, a pre-welded sleeve is a pre-printed film web which is formed into a tube shape (with the printed image on the inside to give total scuff resistance) and with a small overlap that is then bonded together. The tube is seamed and welded to prepare it for the sleeve application machine (see the section on Seaming and Welding).

Sleeves are produced from specially formulated films, which are capable of shrinking biaxially. The main direction of shrink is in the transverse direction (TD) (Figure 7.3). The application of heat causes them to tighten (shrink) around the container.

To label the product the tube is cut to the correct length and opened up for slipping over the pack, from where it is shrunk to a tight fit by the application of hot air, radiant heat or steam in a shrink tunnel placed over the conveyor line (Figure 7.4).

Shrink sleeving is popular for 360 degree labeling on irregular shape packs but can perform a variety of useful roles as a tamper evident neck seal, or a banding device for special offer promotions (Figure 7.5).

THE SHRINK SLEEVE MANUFACTURING PROCESS
An overview schematic of the shrink sleeve manufacturing/application process can be seen in Figure 7.6. The reels of shrink film are typically reverse printed on clear materials using a variety of printing processes. The material is slit to an overlap width

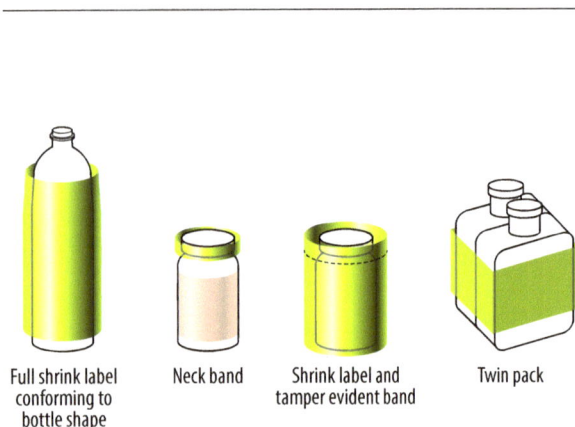

Figure 7.5 Typical shrink sleeve applications

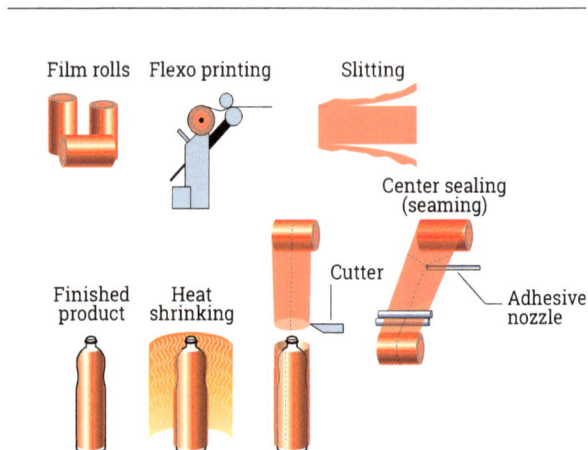

Figure 7.6 Schematic of the shrink sleeve manufacturing and application process

ensuring that the edges of the material are uniform. The film is then seamed by applying an adhesive to the overlap area so as to form a tube, inspected and applied over the container before being shrunk in a heat tunnel.

Each of the stages from material to application will be explained in more detail below.

SHRINK SLEEVE MATERIALS

The type of films used for shrink sleeve production is called transverse (TD) or across-the-web oriented, which are capable of shrinking biaxially to form a tight fit under the application of heat.

Three main base resins are used for shrink films: polyvinyl chloride (PVC), polyester, glycol modified (PETG), and oriented polystyrene (OPS). Some polyester (PET) is also being used.

PVC, which is the predominant resin used in shrink sleeve materials has met with some disapproval amongst environmentalists, as there are some concerns that the burning of PVC films emits chlorine gas into the atmosphere.

The debate surrounding PVC has lead to the growth of alternative materials, such as PETG, styrene (OPS & SBS) and PET. PLA is an environmentally friendly material and is certified as compostable.

In addition to assuaging the environmental concerns of the public, these films can offer other benefits.

Blended polyolefin based films with a low specific gravity of less than one percent are often used ensuring that the sleeve separates from the bottle during recycling (water flotation).

Figure 7.7 summarizes the major film types and structures used for sleeving applications.

PRINTING

Shrink sleeves are typically manufactured from clear, single film which is predominantly reverse printed.

If opacity is required for shrink sleeves the converter would reverse print the design and then add the overall opaque white background as part of the normal printing operation. The benefits of reverse (under surface) printing are the protection of the printed image and the enhanced reflectiveness of metallic printed inks.

A variety of printing processes are used for the print of shrink sleeves, but the majority of long runs are gravure-printed (i.e. typically >500,000 sleeves). However, whatever the sleeve printing process, and whether it is wide- or narrow-web, it is important to have an understanding of the inks used for sleeves. If

Film Resin	Description	Film Structure for Heat Shrink Labels
PVC	Rigid Polyvinyl Chloride	mono layer with printable and sealable surfaces
PS	Polystryrene	mono layer or three layers with printable and sealable surfaces
PET-G	Polyethylenterephthalate Copolyester	mono layer or three layers with printable and sealable surfaces
PETG/OPS	Polyethylenterephthalate Copolyester + Biaxially Oriented Polystyrene Sheet	development in co-extruded materials
PLA	Polylactide	mono layer with printable and sealable surfaces
COC	Cyclic Olefin Copolymer	three layers, with polyolefin core layer and COC surface layers
PP	Polypropylene	three layers with printable and sealable surfaces

Figure 7.7 Major film types and structures used for sleeving

the right pigment selection is not used, then the hot air blowing on the ink can change the ink color as it goes through the shrink tunnel.

It is also necessary to select resin materials that are going to provide good adhesion to the shrink films. Resins need to be very flexible and able to shrink in a similar way to the profile of the shrink films, whether PETGs, PVCs or other substrates. The ink chemistry needs to be compatible with the film selected and adhere to the film as it shrinks in the shrink tunnel.

With today's advances in narrow and medium web printing technology, processes such as digital, flexo and offset are increasingly being used for shorter run lengths, as users demand shorter lead times and more frequent design changes.

A summary of the printing processes used in the manufacture of shrink sleeves along with their advantages and disadvantages is provided below.

GRAVURE

Gravure printing is an excellent method for printing smooth films and provides excellent color

reproduction consistency, high speed and productivity.

It is a process that is ideal for reproducing demanding artwork and photographic elements. The major drawback of gravure printing is the fact that it is more expensive when compared with flexographic printing. The need to produce expensive cylinders makes gravure printing viable only for longer runs.

FLEXO

Flexographic printing in general is making steady inroads into the shrink sleeve market. The increased use of the technology is being driven by the need for quicker processing and reduced costs.

Major advances in terms of the quality of digital image transfer systems, inks and anilox rollers, combined with developments such as larger combination presses or laser-engraved plate technology, have increased its share of the shrink sleeve market.

Other innovations include the development of servo driven presses with individual drives which are more efficient and flexible in terms of substrate handling and the speed of changeovers.

Figure 7.8 Shrink sleeve labels printed using HP Indigo WS6000 digital press. Source: La Catrina Leverages Digital Print

UV FLEXO

UV flexo has a number of advantages over rival processes such as conventional water-based inks, particularly its ink key qualities, print resolution and productivity.

The uptake of UV flexo has risen due to the need to reduce solvent emissions, as required by European and national legislation.

DIGITAL PRINTING

Digital printing is used mainly for lower volumes, proofing and prototyping, but as in other markets it is gaining market share as the demand for shorter runs and rapid turnaround's increases.

Digital printing is ideally suited for mass customization, variable data, multi-languages and product promotions (Figure 7.8).

PRESS CONFIGURATIONS

There are two primary printing press configurations that are particularly suited to shrink sleeve printing.

IN-LINE PRINTING

The in-line printing associated with gravure and narrow web presses can give more flexibility with the types of printing processes used, but it requires sophisticated electronic press registration systems and careful control of heat and web tension.

CI PRINTING

One of the main advantages of central impression (CI) press configurations is the maintaining of very close register on thin filmic substrates. The substrate is wrapped around a large central impression cylinder thus minimizing the movement and material stretch that can take place.

WIDE VERSUS NARROW WEB WIDTH PRESSES

The majority of sleeve printing is carried out using wide web printing formats (more than 1.5 meters).

Narrow web presses (less than 600mm) particularly those with presses adapted to filmic handling using servo drives, UV curing and chilled rollers are increasingly suited to filmic sleeve printing.

PRESS HANDLING CONSIDERATIONS

Shrink film is also more temperamental than other substrates used for labeling, especially when it comes to heat management.

There is a technique to printing film because (converting) equipment will generate heat. Even on press, care is required to avoid shrinking the film. A one percent shrinkage in the film can prevent the label from sliding over the bottle in shrink sleeving applications.

A controlled temperature environment and heat management on press is very important.

PVC and PET, despite their higher percent shrink memory, are considered more 'converter friendly' than OPP. Heavier film gauges also assist conversion.

Shrinkable OPP, despite its lower percentage

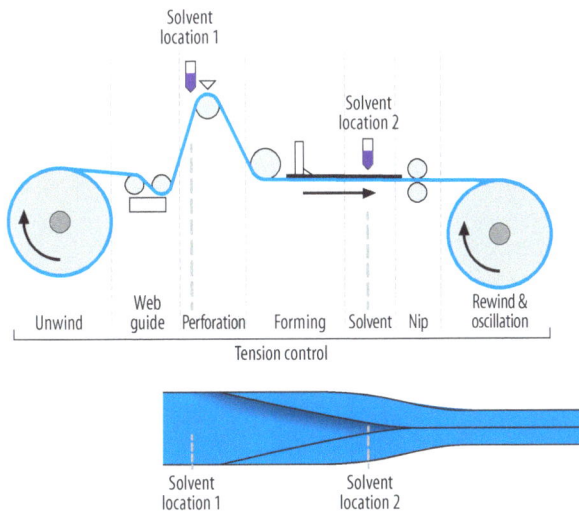

Figure 7.9 The seaming process

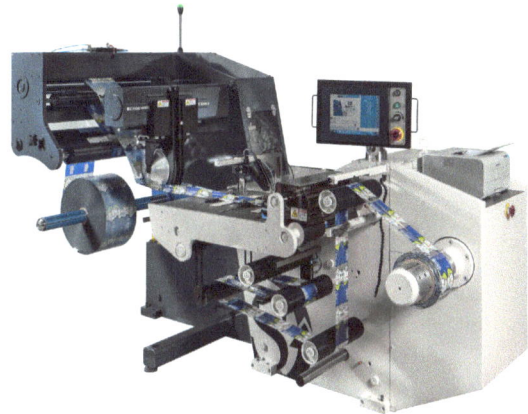

Figure 7.10 Shrink sleeve seamer capable of running at 600 meters per minute. Source: Accraply

memory, requires more care when being converted. This can lead to slower press speeds. However, the leading converters involved in shrink sleeve manufacture are highly skilled and this normally presents no significant problems.

SHRINK SLEEVE SEAMING/WELDING PROCESS

Once the sleeving material has been printed, it needs to be slit and formed into a tube. Solvent is applied continuously and at high speed to the overlap edge on the welder.

The advice now is not to actually crease the film. It creates a memory which becomes difficult to iron out prior to application and creates a line on the decorated container. The aim is to have a U fold rather than a V fold (see Figure 7.11).

A big problem in the sleeve industry has arisen because the film material is not being properly slit. It is a heat sensitive material and If it is slit with the wrong slitting system or a dull blade, it tends to put burrs on the film edge and this will have an impact on the film and subsequent seaming and application process.

Figure 7.9 illustrates the seaming process featuring

typical locations for solvent application. During this process the printed web is perforated at the required sleeve length and folded around a devise known as a 'shoe'.

As the edges of the tube are nipped together the solvent reacts with the surface of the material allowing the surfaces to 'meld' together forming a permanent and secure seal.

The weld areas are generally kept free of print to enhance the bond.

Figure 7.10 shows a picture of an automatic shrink sleeve seamer.

Figure 7.11 Highlights the key terminology used when discussing the seaming or welding processes and clearly shows the seam width, location and overlap requirements for this process.

Seaming speeds and run times will vary according to the sophistication of the seaming equipment used, but turret rewind and unwinds will significantly reduce the running times.

It is always recommended that an inspection takes place after the seaming process has been completed.

SHRINK SLEEVING APPLICATION METHODS

Sleeve labels can be applied by hand or using automatic equipment.

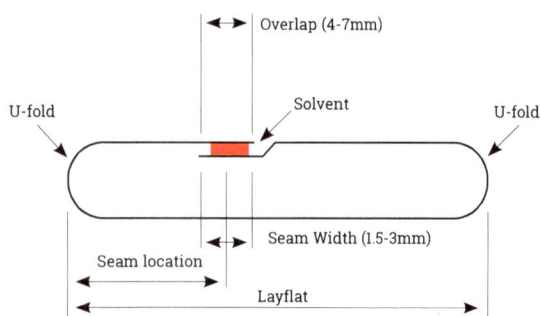

Figure 7.11 Sleeve welding key terminology and specification guidance

During shrink label application the pack or container body is covered or partially covered by the sleeve, then shrunk to fit (Figure 7.4 earlier in the Chapter clearly illustrates the principles).

Shrink applications can be categorized as follows;

- Tamper evident application – where a shrink band is applied to the neck of a container to produced a tamper evident seal
- Pre-forms – a partially shrunk/pre-shaped sleeve is used typically for tamper evident applications.
- Shoulder or part body sleeving – where a sleeve partially covers the shoulder or part of the container walls
- Full body decoration – total coverage of container walls
- Full body sleeve with in-built tamper-evidence – the body sleeve extends over the neck of the container to create a tamper evident sleeve
- Promotional decoration – where perhaps an additional gift or promotional item is shrunk to a product
- Multi-packs – where a number of items are collated and shrunk together to form a multi-pack

There are a variety of both manual and automatic sleeve application systems on offer that can cater for all of the above applications. Sleeving systems

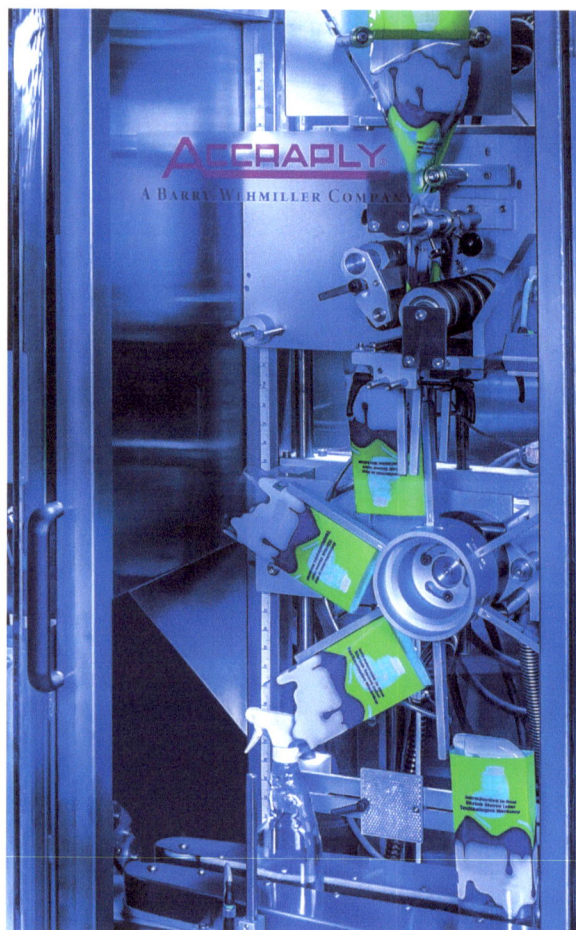

Figure 7.12 Rotary style shrink sleeve applicator. Source: Accraply

typically incorporate the sleeve applicator, shrink tunnel or shrink system (and product handling equipment). Each of the main application systems will be dealt with here.

Rotary-style applicators (Figure 7.12) offer specific advantages for multipacks and odd-shaped containers. Film is registered electronically and fed through a film guide that transforms the lay-flat into a square or rectangle for easier application. The film is then sent through a guillotine-style cutter creating a cut-piece. The cut-piece is met with the rotary tooling

Figure 7.13 Direct apply shrink sleeve applicator.
Source: Accraply

Figure 7.14 A mandrel style shrink sleeve applicator.
Source: Accraply

and held by vacuum. As the tooling rotates, the cut-piece is introduced to the product running on the conveyor below.

These systems often have a lower capital cost compared to mandrel-style applicators. The rotary-style are designed to run at low-to-intermediate speeds and are most suitable for frequent change-over operations, due to the relatively lower cost of change parts. This style is also more tolerant of lower-quality sleeve material.

Direct-apply machines (Figure 7.13) offer a compact design and are generally low-tech and low-cost systems. They are limited to low-to-intermediate speeds and are often suited for large containers.

The film is fed through a guide similar to one found on a rotary-style applicator. It passes through a guillotine-style cutter where the cut-piece is then draped over the conveyor and introduced to the product passing underneath. Direct-apply machines tend to have few electrical components, so cut-length is generally determined mechanically.

Mandrel-style applicators (Figure 7.14) also known as vertical or bullet applicators, are the most sophisticated of the three styles. They are most

suitable for round or cylindrical containers and typically run at intermediate to high speeds. These are often chosen for continuous, 24/7 duties, and dedicated lines.

Film is registered electronically and fed over the film guide fin and mandrel by the film drive rollers. These rollers send the film down to the rotary cutter creating a cut-piece. The cut-piece is forced onto the product passing on the conveyor by the film-drive rollers pulling film into the rotary cutter. They are more demanding of quality materials and have more expensive change parts.

Figures 7.15 and 7.16 show how the shrink sleeve film is guided, registered and fed over a mandrel into a rotary-style cutter.

There is a trend towards applicators capable of applying all label formats (self-adhesive, sleeves, ROSO, and wraparound) on a single machine.

Figure 7.15 Shrink sleeve is guided and registered onto the shrink sleeve applicator. Source: Accraply

Figure 7.16 Close up showing how shrink sleeve film is fed over a mandrel and into a rotary style cutter. Source: Accraply

SHRINK TUNNELS

Shrink sleeves are designed to shrink when exposed to heat and historically the choice of shrink system was dependent upon on a number of factors, such as:

- Whether the container was empty or pre-filled
- Container material - glass or plastic
- Type of product being handled e.g. are the products heat sensitive
- Shrink film being used
- Energy source available
- Line speed

New developments in shrink tunnel technology mean that these factors are less critical today.

Shrink tunnels, used in shrink sleeving tend to operate using hot-air, although infrared and steam are increasingly being used. Their successful performance is primarily a function of the speed and effectiveness with which they can transmit sufficient heat to the film to initiate the shrink action. As exposure of the film in the tunnel is only a few seconds, the heat has no effect on the product; the hot-air only affects the film, causing it to shrink snugly around the product.

Different shrink films have different shrink characteristics, so shrink tunnels need to have a degree of adjustability. Most consist of a conveyor, a heat chamber and a hot-air recirculating system. There are many special types of tunnels available, with some having cooling chambers or pre-heating sections.

Figure 7.17 shows a hot air double tunnel for full body or tamper evident shrink sleeves.

Metallic inks used on some sleeving applications can affect heat transfer and lead to irregular film distortion. Different printed colors too can impact on film shrinkage, i.e. certain colors absorb more heat and therefore lead to an irregular shrink.

Heat shadows caused by the container contours can cause uneven shrinking or distortion of the film and must be considered when designing the shrink system.

In most instances the use of steam shrinking can overcome many of these issues. There are however significant cost issues to consider when providing and managing steam sources. Figures 7.18 and 7.19 show two examples of steam shrink tunnels.

It is worth noting that for some applications such

Figure 7.17 Hot air tunnel (open). Source: Accraply

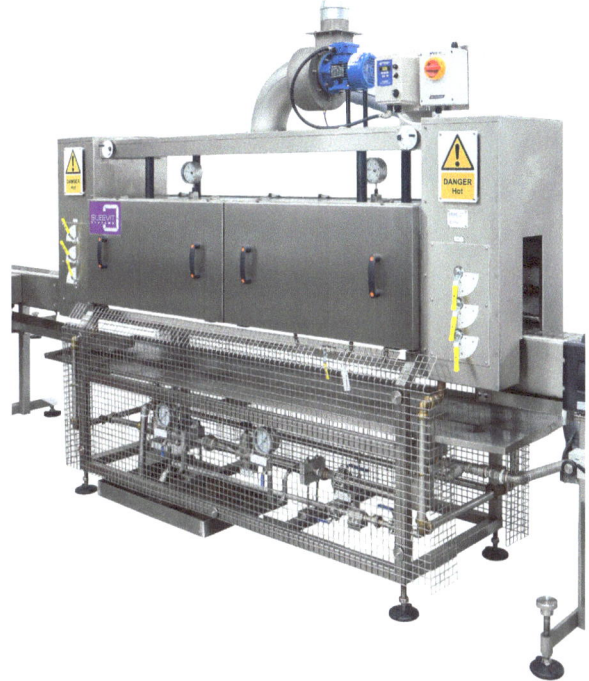

Figure 7.18 Steam shrink tunnel. Source: Accraply

as low speed, tamper evident sealing heat guns or shell reflectors may suffice as a simple shrink method.

PRE-PRESS ISSUES

There are certain key characteristic with shrink sleeves that need to be considered when designing a shrink sleeve.

The drawings in Figure 7.20 highlight the key dimensions, terminology and parameters to be considered when designing a sleeve label. It is important to note that overlap areas will need to be kept free of artwork.

Printing on shrink film presents a number of challenges not encountered when printing on paper. For one, the way the label comes out on press is not the way it will look on the bottle. This is because the label will shrink to the contours of the container and distortion therefore will occur to the printed graphics. It is typical to avoid branding in high shrink areas of the sleeve or too close to the welded seam in order to avoid image distortion issues.

Consequently, calculating distortion correctly is paramount to a successful sleeve label. Traditionally the process was quite laborious. Converters would print a grid pattern on a film, sleeve it, put it on a container and put it through a shrink tunnel, or hit it

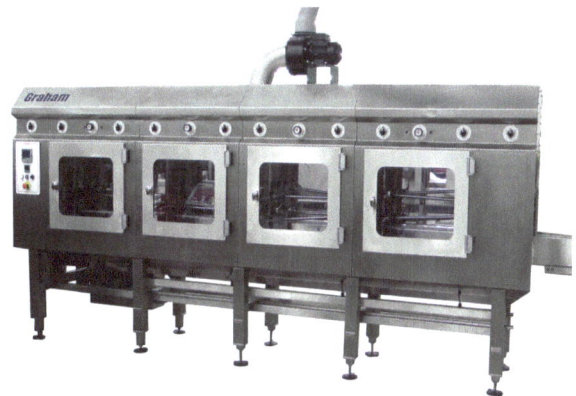

Figure 7.19 Modular steam tunnel with multiple zones. Source: Accraply

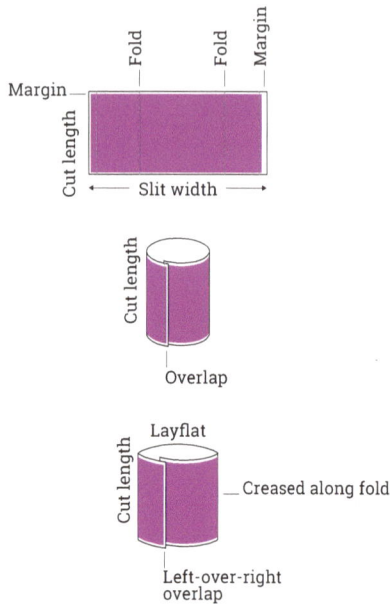

Figure 7.20 Important shrink sleeve design parameters.
Source: Esko

Figure 7.21 Typical roll-fed MD shrink sleeving application
- Pepsi Co Gatorade Thirst Quencher and Gatorade

with a high temperature heat gun to get it to shrink. Then they will cut that off and measure it.

From there the converter would go to a computer program and output the graphics, produce the sleeve, shrink it down, and make sure the graphics appear correctly after shrinking.

Today, there are many software solutions available that can simplify the shrink sleeve workflow process. In general, with these software programs, pre-press can be divided into three different steps or stages. The first step is to prepare the graphics. For that purpose, there are now software suites that conveniently plugs into Adobe Illustrator, so providing a low learning curve. This means that an operator can embrace the technology very quickly.

Once the shape has been created, the next step is to move to a studio designer program in which a virtual sleeve can be applied around the 3D object. This is where the typology, the 3D landscape, can be determined.

The final step is to add the graphics to the 3D shape using, say, Adobe Illustrator. This enables the operator to visualize the decorated container, and even spin it around. It is also now possible to send that file to the customer. There are free apps for iPhone where the customer open up the file on their iPhone, iPad or Android, or Windows mobile device and can view it. With their finger, they can even spin the design around, zoom in and zoom out.

REEL-FED WRAPAROUND LABELING
Other systems of labeling have emerged that can label containers without needing to use a pre-welded sleeve label.

The concept of reel-fed labeling was first introduced by B & H Manufacturing of Ceres, California in 1969 to cater for the high speed labeling

ROSO Production Process

Film → Gravure, Flexo, Offset, Digital Printing → Slitting → Inspection → Seaming on Applicator / UV Adhesive, Laser, Ultrasonic seaming → Auto Application → Shrink Tunnel

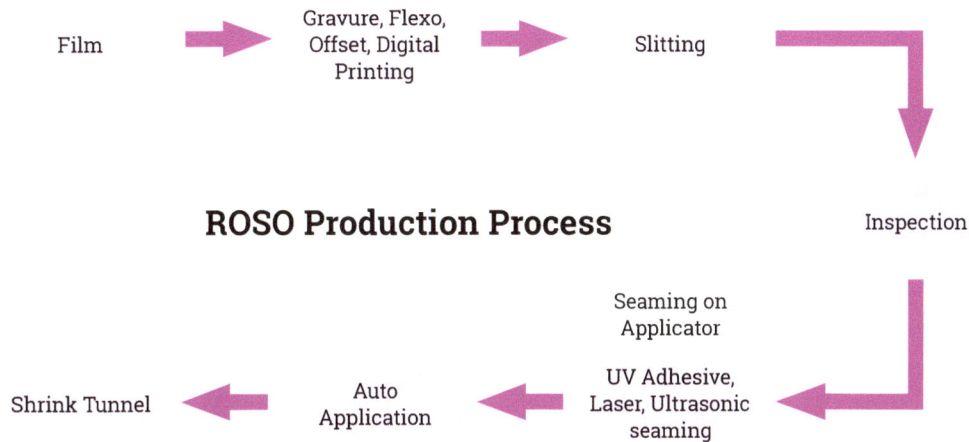

Figure 7.22 Schematic of the ROSO manufacturing/application process

of plastic containers.

These systems originally applied labels of paper or paper-PP laminates mainly to parallel sided containers.

The growth in usage of PET bottles in preference to glass, particularly for carbonated liquids, created a need for compatible label materials that would stretch with the pack and not wrinkle or tear. This trend has driven the use in filmic materials, particularly for carbonated soft drinks and beers market (see Figure 7.21).

For wraparound labeling equipment is designed to use continuous rolls of materials, rather than pre-cut magazine fed labels or sleeves.

With reel-fed systems the cutting, gluing or welding operation takes place on the applicator.

ROLL FED SHRINK (RFS) AND ROSO TM (ROLL ON SHRINK ON) LABELING

In more recent times development in roll-fed shrink labeling has gathered pace.

The descriptions RFS (Roll Fed Shrink Labeling) and ROSO tm (Roll on Shrink On) are typically used to describe roll fed single ply mono-directional shrink

materials used to create shrink sleeves. Labels are supplied in a flat roll format using materials that shrink in the machine direction rather than in the transverse direction (as is the case with pre-welded sleeves).

RFS labels use a metal mandrel on the labeling machine to form the seam, with the seam completed using laser or sonic welding or a heated bar. The completed tube is placed over the container before being shrunk into place.

ROSO mono directional (MD) shrink labels describe roll fed single ply materials that are formed into tubes using adhesives and the container as the mandrel during the application process

Typically ROSO labels have low shrinkage rates of 12-20 percent compared with RFS labels that have higher shrinkage rates of up to 55 percent.

TYPICAL MANUFACTURING AND APPLICATION PROCESS

The process begins with a flat reel of material being printed using one of the major conventional and digital printing processes. On some materials polymer

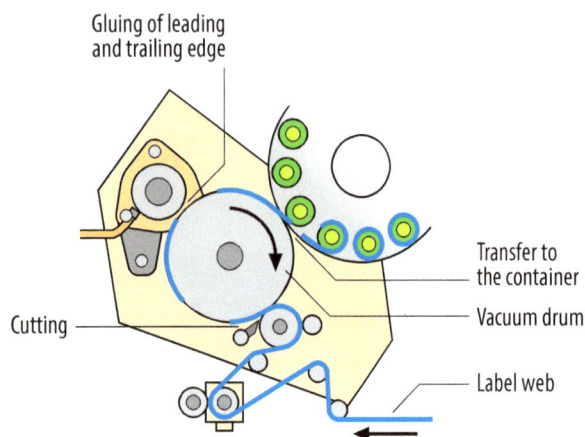

Figure 7.23 Schematic of the ROSO application process

surface modification or coating may be required to improve printability.

The reel is slit to the correct width, inspected for quality and supplied to the end-user or filler for application.

With wraparound style applications the onus for seaming and the application onto the pack is passed onto the end-user.

Seaming and automatic application onto the pack takes place prior to the label being shrunk onto the container. Figure 7.22 highlights a typical manufacturing and application process for reel-fed shrink sleeving.

For non-shrink applications labels are wrapped around the container and typically sealed with hot-melt adhesive.

Whilst acceptable for standard wrap round labeling applications, hot melt adhesives are unsuitable for ROSO applications. The hot air shrink process can re-melt the adhesive thus destroying the seam integrity.

The use of UV adhesive has however partially solved this problem although some improvements are required in order to resist the high shrink forces of steam methods. There are two versions of UV curable adhesive, one of which is water washable to aid removal/recyclability.

ROSO systems today also use other methods for seaming on the applicator such ultrasonic, laser, heat bars and solvent welding.

The label is cut to size prior to adhesive being applied to the leading and trailing edge of the label and then applied to the container (see schematic of the application process in Figure 7.23).

Material is always square cut in order for a positive and secure overlap to be made at speed.

With lower shrink capabilities labeling is generally suitable for decorating parallel sides or areas of the pack with minimal contour deviations.

The containers then pass through a shrink tunnel, which shrinks the film to the contours of the container. Heat in the shrink tunnel need only be applied to the areas where shrinkage is required.

APPLICATION DEVELOPMENTS

Major labeling equipment companies, such as Krones, Sacmi and Trine, have developed roll-fed applicators that incorporate solvent seaming (rather than adhesive) in the trailing edge overlap seam.

Sidel has developed glueless labeling equipment that utilizes a heat bar to weld the label overlap. This process allows the use of higher shrinkage roll-fed shrink films, due to the greater strength of the welded seams.

The effect of container profile on the decoration method.

The container profile and the required conformity of its decoration has the greatest effect on the choice of system and the labeling film that is utilized.

Complex container shapes requiring consistent substrate 'shrinkability' will currently only be successfully decorated using pre-welded shrink sleeves.

Many complex container shapes require film shrinkage factors of 40-70 percent (see illustrations in Figure 7.24).

Developments in applicator technology combined with new film developments will allow ROSO labeling to increasingly penetrate this sector.

ROSO is currently confined to those applications where less than 20 percent film shrink-ability is required i.e. simple container profiles (see illustrations in Figure 7.24).

Complex Container Profiles Requiring
High Shrink Films

Simple Container Profiles Requiring
Low Shrink Films

Figure 7.24 Effect of container profile on the film selection

MATERIALS

Materials used for wraparound systems are supplied in flat web format and are significantly cheaper than those used for pre-welded sleeves.

It is believed that the differential between reel-fed and shrink sleeve films may be due to the increased volumes/yields for reel fed MD ROSO films and/or the method of building memory into MD ROSO films which may be less costly than the TD films used for sleeves.

Opaque and transparent materials used in typical wraparound systems (RFS/ROSO) include;
- OPP
- OPP/PE co-extrusions
- OPS
- Foamed OPS
- PVC
- PET-G

These materials are supplied in 40, 45 and 50 micron thicknesses, with little shrinkage in the transverse direction.

With clear films the labels can be reverse-side printed. White films provide an alternative to paper labels, offering improved gloss, excellent dispensing and better scuff and humidity resistance.

If a design or copy falls within the shrink areas then allowance needs to be made in the design and printing for distortion.

It is important to note that the type of labeler and seaming method being employed will determine the optimum polymer to use for each application.

SHRINK TUNNELS
The main methods of shrinking used for reel-fed wraparound labeling are detailed below.

STEAM SHRINK
Steam is increasingly being used on shrink sleeve applications for the consistency of shrink achieved.

Steam however is not suited to ROSO PP materials. PP needs a temperature of more than 100 degrees to start shrinking, thus it is impossible to use in steam application.

Advantages of steam shrinking includes;
- Energy saving versus hot air
- Uniformity of heating (like a 'sauna'), no need for rotation of the product.
- Easier to switch from one product to another (no redirection of hot air flow)
- Result is perfect at any shrink ratio.

HOT AIR SHRINK
When selecting hot air tunnels as the method of shrinking there are a number of key factors to consider.

HIGH ENERGY CONSUMPTION
Films above 18-20 percent shrinkage can be problematic due to the fans (inside the tunnel) a rotation is required (two conveyer belts at different speed) in order that the product gets heat uniformly.

In addition, the operator will also need to adjust the orientation of the fans inside the tunnel.

BOTTLE DRYING

In many applications containers that are to be decorated need to be completely dry before sleeving.

Sleeved containers that have been steam shrunk need to be dried prior to palletization and ROSO labels require an absolutely dry bottle to ensure blemish free application.

Investment in drying equipment is therefore an additional cost factor to be taken into account.

RECYCLING

A water flotation process is typically used to reclaim film label and container.

Plastic containers are flaked along with their filmic decoration (shrink sleeve or ROSO). The flakes are put through a flotation system which incorporates caustic soda, and which separates the film flakes from the container flakes. Essentially the PET flakes sink and the filmic label flakes float. The PET flakes are recycled, sometimes for use in producing the exterior of new PET containers, and in there are attempts now being made to even produce film from these flakes.

Label materials with a low specific gravity are suited to this method of separation.

PVC sleeved PET containers are problematic due to cross contamination issues.

One interesting environmental issue centers around the removal of the printing ink slurry which is produced during the melting and waste recycling process. The removal of ink slurry to ensure there is no contamination taking place can be problematic.

PROS AND CONS OF REEL-FED SYSTEMS

Some of the key advantages and disadvantages of roll-fed systems are highlighted here.

Advantages
- Reel-fed applications are generally lower cost versus pre-welded sleeves
- Supplied in a flat web allowing more labels per reel
- Process is relatively simple as it only entails printing and slitting
- No pre-seaming of sleeves required (including seam inspection)
- Possible to seam without adhesives and

solvents therefore easier to recycle
- Non-adhesive based welding eliminates expensive UV lamps and difficulties associated with use of UV adhesives
- Higher labeling speeds possible

Disadvantages
- Shrinkage factors tend to be less than welded sleeve formats – limited shrinkage in transverse direction
- Less scope for full body coverage on difficult shaped container profiles
- High capital cost attached to reel-fed applicators, with expensive change parts and time consuming changeovers between bottle shapes and sizes
- The inclusion of tamper evident features is very difficult for ROSO applications.

NEW DEVELOPMENTS
Adhesleeve – PE Labelers

This is a new development in pre-adhesive self-wound shrink labels from Italian applicator specialist PE Labelers.

PE Labelers has been in the labeling industry for 12 years and supply parts for Sidel and Ocme machinery.

Called 'Adhesleeve', the system uses a pre-applied clear adhesive strip on the back of the ROSO label. The adhesive is applied at the printing stage and cured in-line, so the reels are supplied to the application line with the graphics and adhesive completed.

The label is cut on-line, wrapped around the container and the trailing edge is then fixed to the adhesive strip creating a sleeve which is positioned on the container. The container then travels through the shrink tunnel and emerges ready for filling.

This new method of application allows the shrink label to be applied to a container without the need for a special hot-melt glue together with UV tunnels.

Eliminating the use of the hot-melt glue implies several advantages: from the absence of fumes produced by heating the hot melt, to the elimination of solvents to clean the application cylinder. Furthermore the label, as it is no longer contaminated by the hot melt, is 100 percent recyclable.

Figure 7.25 Examples of bottles decorated with high stretch sleeves (Triple S). Source: Labels & Labeling

The development of the Adhesleeve project has lead to the realization of a new method of label cutting, no longer affected by the interaction of fixed and rotary cutting blades.

Thanks to this new cutting system the adjustments to the cutting unit in case of various film thickness, changes in the ambient temperature and worn out blades are no longer necessary and the replacement of the cutting devices, which do not require any servicing, is effected without tools. All this allows a 100 percent elimination of production stops, thus improving the labeling efficiency.

(Note: it is speculated that the cutting method could be air knife or laser).

Sleeving summary

The popularity of shrink labeling shows no sign of abating, with double digit growth rates being achieved.

The penetration of ROSO labeling has suffered in the past through its inability to offer a solution to the decoration of containers with complex profiles/contours requiring high shrink MD films.

Developments in servo press technology and in UV flexo printing is encouraging narrow web converters to enter the shrink label market particularly for shorter runs.

The conversion costs attached to the seaming of shrink sleeves can significantly add to costs versus ROSO.

With ROSO technology the onus for seaming is passed to the end user. UV adhesives used for ROSO seaming are expensive and can be messy, however developments in welding technology are removing this barrier.

There are a number of new development films appearing both for ROSO and indeed sleeves (including PLA, COC, PET/OPS/PET).

The expansion of ROSO labeling for long run brands will be accelerated by the further development of high shrink blends.

A key driver for ROSO labeling remains the favourable raw material cost differential versus shrink sleeves and the higher yields it can achieve.

An interesting development is the Adhesleeve system from PE Labelers that eliminates seaming and glue problems using ROSO labels with a pre-applied adhesive strip.

Cut and stack film labels

Film labels when specially treated for static can be applied as cut singles by magazine fed applicators such as the Krones Canmatic.

Stretch sleeving

Stretch sleeves produced in low density polyethylene are designed to be smaller in circumference than the object bottle and are subsequently stretched during application.

The sleeve is securely held in place by the elastic properties of the film alone, therefore eliminating the need for heat or glue during application.

Stretch sleeves have had limited appeal and were seen primarily on large bottles of fruit juice and low quality, large container applications (motor oil) (Figure 7.25). Improvements in stretch sleeve materials and application processes are widening the appeal of this decoration method.

Key benefits

Heat shrink sleeving methods typically use an oversize shrinkable film tube, which will then be shrunk to a tight fit using a shrink tunnel. The latest stretch sleeve technology however takes an LDPE printed film tube and stretches it over the bottle or container and achieves a tight fit without any need for an energy-intensive shrink tunnel.

Stretch sleeving therefore uses less film and requires no shrink tunnel with significant potential cost saving and environmental benefits.

As the name implies stretch film will expand and contract with the bottle and is ideal for carbonated beverages and squeezable bottles that need to flex.

Not being heat sensitive, the stretch sleeving operation can be carried out on hot-filled containers.

LDPE stretch film used has a low density (0.91 gr/ccm) allowing easy and cleaner separation, thereby supporting PET and bottle-to-bottle recycling.

Chapter 8

Flexible packaging

Flexible packaging has been growing rapidly in recent years and replacing other forms of packaging, such as rigid plastics containers or corrugated materials.

Pre-printed packaging in particular has many distinct advantages of other decorative packaging. In effect the wrapper itself acts as both the packaging and the label, and so requires no further decorative input. It is therefore low cost, light and disposable and is an extremely attractive option for brand owners.

However, where the packaging itself comes into contact with food products, either directly or indirectly, there are numerous rules and regulations in place to protect the consumer and to ensure that inks, adhesives and other materials do not migrate into the product, or contaminate them in any way. These issues will be highlighted in this Chapter.

The Chapter will also explore a number of the main flexible packaging systems in use and their role as a carrier of product decoration and branding, in a variety of market sectors.

MARKETS FOR FLEXIBLE PACKAGING

The largest sector of the flexible packaging market are for applications in food and retail, but other non-food sectors are growing fast (Figure 8.1).

Major applications for flexible packaging are diverse and include the following;
- Multi-colored frozen food packs
- Sweet wrappers
- Confectionary packs – e.g. Mars wrappers etc.
- Garden centre and horticultural packs – fertilizer, bark, grass seed etc.
- DIY packs – wallpaper paste, plaster etc
- Stand-up pouches – washing liquids, detergents, soups etc.

Figure 8.1 Examples of printed filmic packaging used in the food sector

End-use markets for flexible packaging

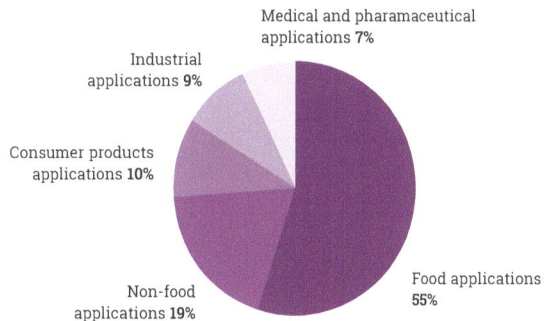

Medical and pharamaceutical applications **7%**

Industrial applications **9%**

Consumer products applications **10%**

Food applications **55%**

Non-food applications **19%**

Figure 8.2 Global flexible packaging market - the key end-use markets for flexible packaging

Print and packaging – flexible packaging

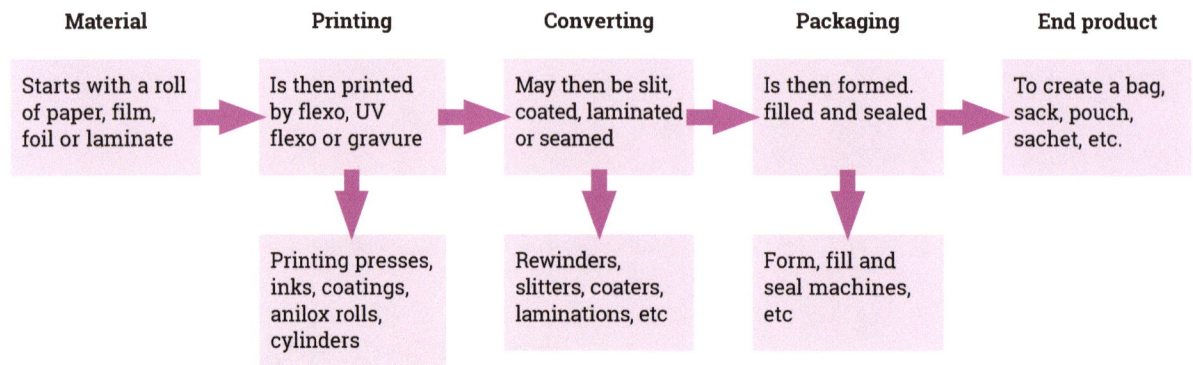

Material	Printing	Converting	Packaging	End product
Starts with a roll of paper, film, foil or laminate	Is then printed by flexo, UV flexo or gravure	May then be slit, coated, laminated or seamed	Is then formed. filled and sealed	To create a bag, sack, pouch, sachet, etc.
	Printing presses, inks, coatings, anilox rolls, cylinders	Rewinders, slitters, coaters, laminations, etc	Form, fill and seal machines, etc	

Figure 8.3 Typical converting process for flexible packaging

- Sachets
- Lidding for yogurt, cream, dessert, pots etc.
- Agricultural packs – feeds, seeds etc.

Some product categories have notably shifted almost entirely to filmic packaging, for example baby food and some tobacco products.

The key end-use markets for flexible packaging are featured in the chart (see Figure 8.2).

THE CONVERTING PROCESS
A typical converting process for flexible packaging from material through to the end-product is illustrated in Figure 8.3 below. Each of these steps will be covered in the paragraphs that follow.

MATERIALS
Flexible packaging uses a combination of materials to provide properties that protect the package contents.

Flexible packaging materials are typically produced from three main materials types:
- Plastics
- Paper or paper-based
- Aluminium foils and foil laminates

Although paper is widely used, it is the growth in the use of filmics that has been most dramatic.

Of these materials, some 75 percent are plastics, of which the main film materials used are polyethylene (PE), low density polyethylene (LDPE) and bi-axially orientated polypropylene (BOPP).

Aluminum foils are used where their better barrier properties give them the edge over flexible films or papers. They are used for confectionery, ready-meals, pharmaceuticals, soups and sauces, preserved foods, liquid foods.

Paper and paper based materials are used for a wide range of printed bags and sacks. Paper face constructions are typically used for dry goods like sugar, seasoning, soup mix and cocoa. Constructions between three to four mils can be surface printed with a varnish. By adding over-lamination, the material provides better stiffness and puncture resistance.

Many materials used for flexible packaging involve

the use of multi-layers which are created using two different laminating processes.

Extrusion laminating is a process in which layers of multilayer packaging materials are laminated to each other by extruding a thin layer of molten synthetic resin, like polyethylene (PE), between the layers.

Adhesive laminating is a process in which individual layers of multilayer packaging materials are laminated to each other with an adhesive (described later in the section on in-line laminations).

There are generally three components in a flexible packaging structure – the exterior, the barrier, and the sealant. The exterior layer is the print surface.

The second component is the barrier layer which provides protection based on the product being packaged, the desired shelf life and the storage and distribution conditions required.

The sealant layer is a material that will adhere to itself, or to another film when heat and pressure are applied to produce hermetic seals that prevent gases from penetrating through the seals into the package.

It is typically applied to the inside layer of a multilayer structure on the side that comes in contact with the product.

PRINTING OF FLEXIBLE PACKAGING
All flexible packaging formats are printed in multiple colors on web-fed (reel printing) machines, including wide-web, mid-web and some narrow-web presses. The printing of films, some often quite thin, means that presses may require sophisticated tension control, perhaps heat management systems, static control, corona treating and registration, etc.

In terms of the printing of flexible packaging, gravure is regarded as being superior in print quality, although the quality of flexo has improved dramatically and is now considered cost-effective compared to gravure.

With more and more installations around the world, many converters are now seeing the value in the production of flexible packaging using digital printing.

Digital printing's ability to react swiftly to market demands and to produce small print runs without the corresponding loss of time that used to accompany the production of the new tooling required for flexo,

offset or rotogravure printing, offers significant and positive benefits.

Traditional flexible packaging run lengths continue to trend downwards as wide web converters struggle to print the smaller run lengths economically. There are however many narrow web in-line style printing presses (servo-driven machines in narrow web widths) that are now able to handle the shorter runs. Servo drives and controls have helped overcome the main challenges that have historically plagued the flexo market, such as gear marking.

Some 60 percent of flexible packaging printing is direct onto the film surface, whilst 40 percent is printed on the reverse and laminated.

Apart from print performance, flexible packaging films may have to withstand lamination, sterilization, pasteurization and thermo sealing and therefore may be surface coated or treated.

Inks and varnishes play a key role in not only providing the decorative appeal, but also adding barrier properties, coatings and seals.

As most flexible packaging is used for food or for sensitive applications (body-care, cigarettes, etc.), it is important that neither the materials nor the inks should be able to contaminate the product by odor, migration or through trace particles. A small, but increasing use of UV inks is tending to push flexibles printing towards flexo (covered in the following section on migration).

Standard tests for flexible packaging may include sniff tests, color assessment and flavor tests.

As there are a number of different foil laminations available, substrate surface preparation is key to achieving high quality print. Most plastics – polyethylene, polypropylene and polyester, have chemically inert and non-porous surfaces. Corona treatment is a popular option used to increase the surface tension of a given material to promote adhesion.

IN-LINE LAMINATIONS
Laminating is the process through which two or more flexible packaging webs are joined together using a bonding agent to improve the appearance and barrier properties of the substrate.

The choice of the most suitable web laminating

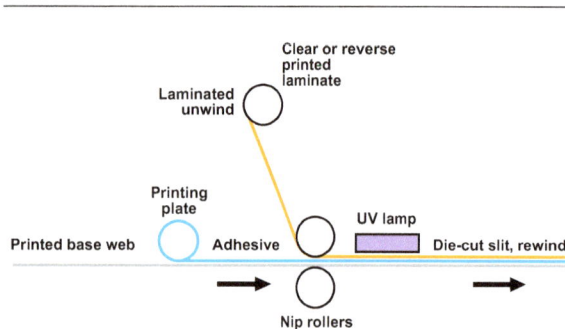

Figure 8.4 - Wet Lamination Process

process is mainly dictated by the end-use of the product. A number of different technologies are available that cover the wide variety of applications in the food and non-food packaging industries.

Laminating machinery can be classified according to the type of bonding agent used to produce the laminates. These types are:

Solventless lamination: The adhesives used do not contain solvents. They dry by chemical reaction and therefore do not requiring a drying system. This method is used widely in flexible packaging since the chemistry is relatively simple and the applications broad.

Wax lamination: The adhesive is a wax or hot melt which is applied in a liquid state to one of the substrates prior to the substrates being brought together. This process allows the production of paper-paper or paper-aluminum foil laminates that are widely used for the packaging of biscuits and bakery products.

Dry lamination: Is the process where the bonding agent, dissolved into a liquid (water or a solvent), is applied to one of the webs, before being evaporated in the drying oven. The adhesive coated web is laminated to the other under strong pressure and using heated rollers, which improves the bond strength of the laminate.

Wet lamination: With wet lamination the adhesive is in a liquid state when the laminate and base substrates are brought together. It is commonly used to produce a paper-aluminum foil laminate that is widely

used in flexible packagingand in the self-adhesive labeling industry. In the process of wet lamination the adhesive is applied to either the reverse side of the film laminate or alternatively to the face of the printed substrate to be laminated. The liquid adhesive is applied by a roller coating system. The two substrates, one of which has been coated with the adhesive, are then nipped to form the bond (see Figure 8.4).

In all of the laminating processes described the resulting laminated web is then rewound into a finished roll.

BARRIER COATINGS

A barrier coating or sealing coat can be applied to prevent migration of ink, adhesive, or other substances through the face material.

Barrier and other functional coatings encompass materials that are coated onto substrates to provide a barrier to protect selected packaged goods. Barrier coatings, providing barriers for food packaging requirements, may include protection against oxygen and aromas, liquid water and water vapor, oils, and grease.

An effective barrier can prevent both losses from the packaged product, and penetration into the package, both of which can affect quality, and shorten product shelf life.

Packaged food products are being maintained fresh longer as a result of new materials, and food processing developments. For example, O^2 scavengers are now being used that work within a sealed package to limit O^2 reaction with a food product. Combined with effective O^2 barrier packaging, food packagers have the ability to improve shelf life, preserve product appearance, and flavor, while minimizing preservative use. Additionally, antimicrobials, while under siege, have been proven effective as additives to coatings, and packaging films, in combating food sourced illnesses.

Nanotechnology is being applied to improve the gas barrier properties of coatings. In doing so, nanoclay is dispersed in barrier coatings, resulting in a platelet orientation that creates a 'torturous path' for gas molecules to traverse, yielding a very thin film, effective gas barrier.

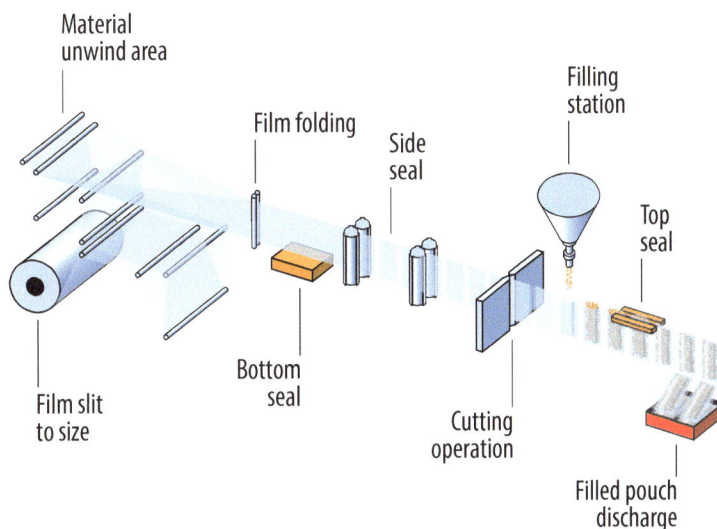

Figure 8.5 A typical pouch forming and filling machine

FORMING THE PACK

The final process in creating a filled and decorated pack requires that the printed film is converted into a bag, pouch, sachet, tube, sack or other shapes using equipment that typically forms, fills and seals the pack. This process will vary depending on the type of pack required. The main types of end-use formats for flexible packaging will be dealt with in the next section.

FLEXIBLE PACKAGING END-USES
Pouches

Printed flexible pouches can be a pre-formed, three-side sealed pouch, or formed as an in-line operation with the filling and sealing combined on a forming and filling line.

A typical pouch forming machine is illustrated in Figure 8.5. This equipment unwinds a pre-printed roll of film, folds and seals the film on the edges, leaving the top open. The pouch is then filled from above with product before being sealed to form the sealed pouch.

There are numerous pouch formats, styles and shapes that are available.

Stand up pouches are a laminated film bag, typically made of plastics or a blend of plastic film and aluminum foil. They can be printed with any color, logo, or design, so the potential to really make an impact on retail shelves is very high.

Stand up pouches are excellent options for both dry food packaging and liquid products.

Made from a continuous web of material, the first step in the stand-up pouch manufacturing process happens when the material is passed through a set of plows that fold a W-shape gusset into the bottom, so it can stand up.

The popular retort pouch is a flexible laminated food package that can withstand thermal processing. The choice of materials for the manufacture of retort pouches is very important. The material must have sound structural integrity and be able to withstand retort temperatures as well as normal handling conditions. A hermetic seal is achieved in retortable pouches by the fusion of two heat-sealable layers (such as polypropylene) to each other.

Single or re-fill pouches in particular, are being used for an increasing range of products, from liquid

Figure 8.6 Digitally printed stand-up pouches with personalization

Figure 8.8 Typical wrapper style packaging (gravure printed)

Figure 8.7 Stand up pouches for the food sector

Figure 8.9 Typical single serve packs for a home improvement range

detergents to beverages and lubricants – and this is expected to continue to grow rapidly. Examples of stand up pouches can be seen in Figures 8.6 and 8.7.

Wrappers

Pre-printed wrappers and decorative bands are an effective way for pack branding and packaging. Wrappings are typically used on confectionary, sweets, gift wrapping, butter packs, etc.

In many cases products are completely over wrapped in paper or film to form an airtight seal. Figure 8.8 is a good example of this style of decorative packaging.

A Form Fill Seal (FFS) process called 'flow wrapping' is used to form a close fitting and air tight bag around the product. A flow wrapping machine is a horizontally operated machine with the packaging material mounted above the operating level. Typically, the product is loaded horizontally with a longitudinal seal formed below the pack.

Over-wrapping is another method of packing a product that does not create an airtight seal, but creates a pack like a neatly wrapped gift.

Both flow wrap and over wrap methods typically

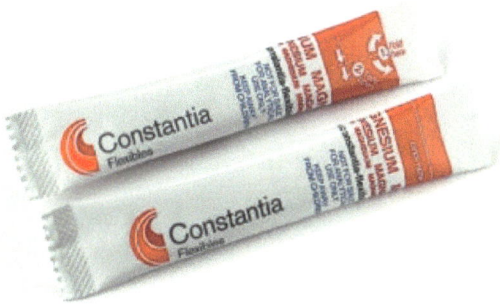

Figure 8.10 Typical stick pack construction (Constantia Flexibles)

Lidding foil

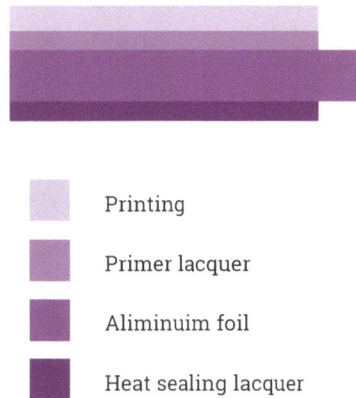

Printing

Primer lacquer

Aliminuim foil

Heat sealing lacquer

Figure 8.11 Structure of a typical aluminium foil lid

use polypropylene film (BOPP) to create the wrap, although other film and paper substrates can also be used.

Cold sealing methods are used, particularly in confectionary applications. The adhesive is coated onto the material. During the wrapping process the material is folded onto itself and sealed via a sealing wheel.

Single use packs
Consumers are finding significant benefits in single serve flexible packs that minimize waste, supports eating on-the-go lifestyles or allow users to sample small quantities of product (see Figure 8.9).

Single use packs are typically smaller versions of stand-up pouches or sachets.

Stick packs are another method of accommodating single use products. Named after its long and slender shape, stick packs are a type of vertical form, fill and seal (FFS) pouching (Figure 8.10)

Lidding
Some containers such as tubs, trays or jars, now have a clear printed plastic film heat sealed onto the container: This is often called a lidding film.

Typical lidding materials include paper, metalized PET or PP. Quite widespread is the use of aluminium foil for lidding, particularly for water cups, dry foods, yogurt pots, ice-cream, etc. A typical lidding film

Figure 8.12 A heatsealable polyester film which peels cleanly from trays in ambient or chilled conditions. (KPeel 3G - KM Packaging Services)

construction is illustrated in Figure 8.11.

Lidding films not only seal and protect the product but can perform an important decorative function. Most lidding films are designed to be peelable to allow easy access to pack contents (Figure 8.12).

Peelable seal lids require a polymer layer on the inside to facilitate the heat sealing.

Figure 8.13 Digitally printed vertical form filled seal, 3-ply laminated stand-up bag for cocoa and chocolate drink powders (Uni Packaging - Bensdorp's Barista range)

Some lidding applications require a heat seal coating applied to the film construction. The coated film passes over a pre-heat station where it warmed before it is sealed to the tray or pack, via a sealing bar or platen set at a desired heat, pressure and dwell time. The sealed tray or pack is then die-cut to shape.

Pot applications such as yogurts use heat seal pre-cut lids or diaphrams matched to the shape of the container.

Form Fill Seal (FFS)

Form-fill-seal (FFS) machines are used to form the package, fill it with a wet or dry product and seal it closed. Most FFS systems use a roll of flexible packaging film which is shaped and sealed to form the primary package, such as a wrapper or pouch (Figure 8.13).

FFS machines can be positioned either vertically or horizontally. Vertical machines form and cut packages. The product is then dropped into the package before final sealing (Figure 8.14). Horizontal machines are used in cases where dropping a fragile product (such as a cake or biscuit) vertically may cause damage and instead the product is placed into the package horizontally (Figure 8.15).

FFS machines are able to fill the flexible pack from either the top or the side. In most cases heat sealers

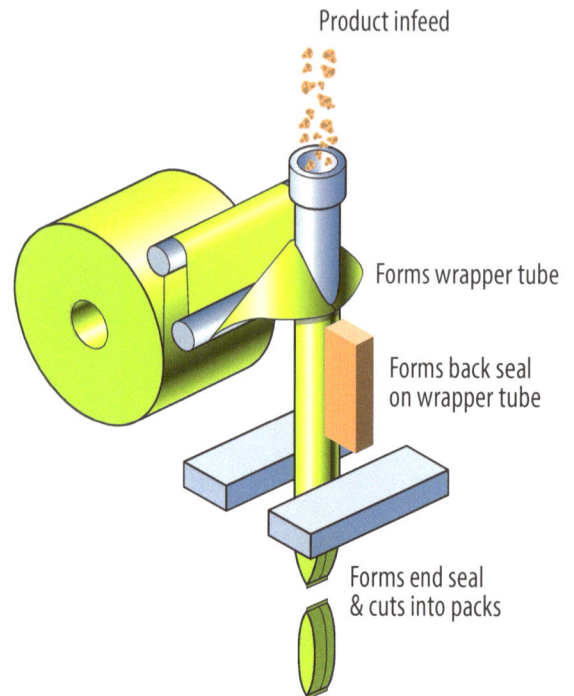

Figure 8.14 Typical vertical form fill seal system

apply heat to the sides of the package and melt the substrate material together to form a seal. Ultrasonic sealing is a new development which tends to be used for heat-sensitive products and permits sealing through liquids.

Sachets

A variety of materials can be used for sachet packing including aluminium foil, paper backed foil and PET foil.

Individual packs can be designed and printed in a range of colors to reinforce branding, as well as displaying all the required regulatory and product information.

Many different sizes of sachets can be filled with a range of products including powders, tablets and capsules or liquid.

Packaging machines are used which take flexible packaging material, to form a package which is then filled and sealed in a sequence of operations to form a

Horizontal wrapping process

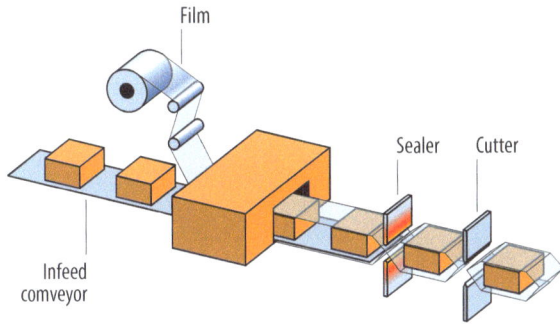

Figure 8.15 Typical horizontal form fill seal system

Figure 8.16 Bottle neck foiling conveys the impression of high quality

three or four side sealed sachet. The process is similar to that illustrated in Figure 8.6.

There are two basic types of sachets, a fin seal type which is a face-to-face seal around the pack and a pillow style which is a crimp seal on the top and bottom edges and a flat seam running down one side.

The majority of machines used for these type of operations, are of the vertical form fill seal type, although horizontal form fill seal equipment is sometimes employed.

ALUMINIUM FOIL BOTTLE CAPS AND NECK FOILS

There are many types of bottle cap liners in use. Whilst they can be printed, their main function is to seal and protect the contents of the pack.

A popular seal liner is the induction seal which contains a foil laminate (known as an inner-seal) which is welded to the top lip of jars and bottles and creates a hermetic, tamper evident seal. Aluminium foil's ability to be a total barrier to light, atmosphere and liquids is the principle reason for its use in caps, capsules and lids.

The sealing process takes place after the filling and capping operation. The capped containers pass underneath an induction sealing system which produces

an electromagnetic current and the foil laminate generates electrical resistance, heating the foil. The hot foil in turn melts a polymer coating on the inner-seal. The heat, coupled with the pressure of the cap, causes the inner-seal to bond to the lip of the container. Heat seal closures are compatible with the wide variety of plastic containers – PS, PP, PE, PVC and PET.

Aluminium neck foiling is used on a large scale particularly for premium brand beers, wines and champagnes to convey the impression of high quality. Unsupported neck foils are often printed and are totally malleable, and can cover curved and shaped areas of the bottle neck and closure, to create a decorative finish and provide evidence of tampering. Neck foils are frequently embossed giving a distinctive appearance (Figure 8.16).

FOOD LABELING APPLICATIONS

In the food sector consumers are increasingly concerned about labels and/or packaging contaminating products. This concern is heightened in instances where packaging materials are in direct contact with food contents, but many of the issues relating to migration apply to all types of labeling and packaging.

Most countries have standards which determine

Figure 8.17 A guide to some of the most important aspects of food labeling regulations and directives

which label and packaging materials may come into direct contact with food-stuffs and human skin. These products are prohibited along with those where there may be some transfer or migration of substances to food. The label user and/or label manufacturer are held liable for any failure on their part to comply with standards.

The labeling requirements for foods are set out in the relevant EU, FDA and national regulations around the world. They generally relate to most prepared foods, such as canned and frozen foods, breads, cereals, snacks, desserts, etc. Nutritional labeling may be voluntary, while other foods may have more detailed separate labeling requirements.

There are a number of directives and regulations governing food labeling, but these are complex and not always easy to understand. The EU started to harmonize legislation on food contact materials several years ago, but fully harmonized legislation does not yet exist for all materials. Much of the basic regulatory background is contained in Regulation (EC) N° 1935/2004 and in particular article 15 of this regulation.

The article lays down common rules for packaging materials which come, or may come, into contact with food, either directly or indirectly. It also seeks to protect human health and consumers' interests throughout the European Economic Area. It covers a wide range of different materials, including all papers and boards, plastics, inks, adhesives and coatings

Any substances which can reasonably be expected to come into contact, or which can transfer their constituents to food are covered by the regulation. It seeks to ensure that the labeling of foods 'shall not mislead the consumer'.

In America the FDA (US Food and Drugs Administration) is the U.S. federal agency responsible for ensuring that foods are safe, wholesome and sanitary; that human and veterinary drugs, biological products, and mechanical devices are safe and effective; that cosmetics are safe; that electronic products that emit radiation are safe. The FDA also ensures that these products are honestly, accurately and informatively represented to the public, including issuing regulations governing the use of self-adhesive labels for contact with foodstuffs (FDA 175.125 for direct contact, and FDA 175.105 for indirect contact).

The U.S. Federal Food, Drug and Cosmetic Act (FFDCA) defines food 'labeling' very broadly. It covers all labels and other written, printed, or graphic matter upon any article or any of its containers or wrappers, or accompanying such article. The term 'accompanying' extends to tags, leaflets, circulars, booklets, brochures, instructions, and even websites.

The Nutrition Labeling and Education Act (NLEA), which amended the FFDCA, requires most foods to show specific nutrition and ingredients on the label. Food, beverage and dietary supplement labels that show nutrient content claims (for example 'low fat' or 'contains vitamin XYZ') and certain health messages

MIGRATION CONSIDERATIONS IN FOOD LABELING

Ink migration

Migration from substrates

Migration from adhesives

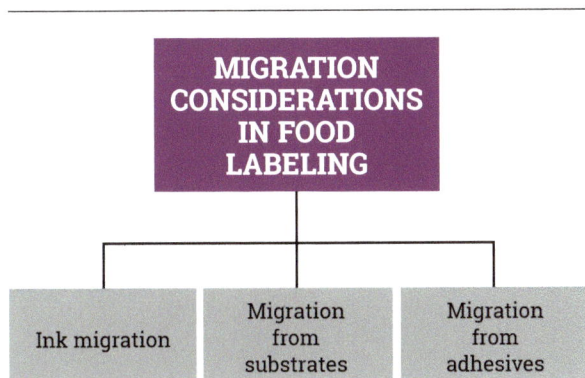

Figure 8.18 Key areas of migration to be considered in food labeling

have to comply with specific legal requirements. Furthermore, the Dietary Supplement Health and Education Act (DSHEA) has amended the FFDCA, in part, by defining 'dietary supplements'. It also adds specific labeling requirements for dietary supplements, and provides for optional labeling statements).

A US government advisory body exists to check compliance of any label with current rules. This service (which must be paid for) is recommended for manufacturers and label/leaflet vendors. A recent regulation (May 2016) changes US requirements regarding health criteria to be mentioned on labels.

Figure 8.17 highlights some of the most important aspects of food labeling regulations and directives.

MIGRATION

There are rules and regulations in place to ensure that inks, adhesives and other materials do not migrate into the product.

For Europe, regulation (EC) No 1935/20041 requires that materials and articles which, in their finished state, are intended to be brought into contact with foodstuffs, must not transfer any components to the packed foodstuff in quantities which could endanger human health, or bring about an unacceptable change in the composition or deterioration in organoleptic properties.

Figure 8.18 highlights some of the key migration issues to be considered when labeling or packaging food.

Although the risk that chemicals may leach through labels and packaging is real, the fear arises through most people's ignorance of what chemicals there are in ink. As a result, label converters are turning pre-emptively to more expensive low-migration inks.

Some ink manufacturers are promoting UV LED inks not only for the energy economy, but also because this drying technology can be better controlled than traditional mercury lamp drying.

There has been some concern about the risk of contamination through mineral oils contained in substrates. Paper materials however are mostly made from virgin fibers, and synthetic substrates have (so far) been less often accused, and are available in 'high barrier' options from several producers.

Adhesives used for self-adhesive labels can also be a potential source of migration and contamination, particularly when these labels are applied directly to foodstuffs. The two most commonly used adhesives for pressure-sensitive labels are hotmelt and acrylic emulsion. Nearly all adhesives stick because they contain resins, and the lower the resin content, the lower the initial tack. This problem is even more acute with filmic labels. However recent developments using a multi-layer adhesive technology have resulted in a virtually resin-free adhesive, which can be used for filmic food labels as well as for moist or fatty surfaces.

Knowledge around barrier protection is crucial for servicing the market. Suppliers must be able to provide the right sealant for a given application and to support clients with 'fitness for use' testing.

CONCLUSION

One of the main factors driving the growth in flexibles is cost savings, particularly when flexibles are compared with labeled or decorated rigid packaging. Users of flexible can claim an improved carbon footprint, because of the weight and volume savings achieved. As mentioned previously migration and food labeling regulations need to be carefully considered in those sectors.

Recycling, environmental friendliness and sustainability have become issues for all packaging, providing the opportunity for new packaging materials development. As a result, continual improvements in polymer films, surface treatments, and coatings are yielding new packaging alternatives while respecting environmental impact.

Film structures are clearly moving to thinner laminations with stronger barriers and improved seal properties, which may require investments in servo-driven presses with special tensioning capabilities. At the same time, packaging machinery is being built to run faster.

In recent years the market for flexibles has been driven by innovative new origination, print and press technologies. There is also a growing realization that narrow-web digital presses also have a significant role to play in meeting manufacturer's demands for shorter runs, lead times, variable data and more product personalization.

Chapter 9

———

Total applied cost

———

As we have seen in the previous Chapters there are many systems and methods that can be employed to decorate a pack or product.

———

In some cases there may be overwhelming reasons why one decoration is selected over another, but inevitably an assessment of costs will form an important part of the decision making process.

It is important to recognize that there is a big difference between the price of a label and the total cost of the labeling technology or system.

Taking a holistic view of the costs associated with different product decoration technologies can significantly influence the selection of a decoration or labeling method.

This Chapter aims to provide the reader with a greater understanding of the concept of 'total applied cost' and will help them to make more informed decisions.

There are many instances where making a straight comparison of label prices is valid, but end users need to look at the whole picture in order to make a full and realistic assessment of the costs of product decoration.

From a supplier's perspective, encouraging end users to widen their view can also open up new opportunities to present a more compelling case for their products.

EXPLAINING THE CONCEPT OF 'TOTAL APPLIED COST'

The total cost of a label is not merely the cost of the material and its conversion…there is a much wider range of factors at play. Customers trying to make comparisons between a wet-glue and a self-adhesive label or a shrink sleeve and an in-mold label, for

example, would be naïve to make a decision based on simple price per thousand comparisons.

In-mold and direct decoration, for example, often require the user to store large quantities of pre-labeled containers on-site, in order to cope with the flows of demand across a number of variants. Storage, inventory and obsolescence are all part of the labeling cost and have to be considered at the very start when decisions are being taken as to how a product will be produced, presented and marketed.

Likewise the cost of application equipment, manning levels and labeling efficiencies are likely to be key components of the cost equation.

The concept of total applied cost encompasses all costs attributed to the labeling process from start to finish. Evaluating different decoration methods using this wider definition of costing can dramatically influence the selection process.

BASIC COSTING

Material content is an important factor in the price per '000 and can comprise up to 70 percent+ of the price of the labels, dependent on the form of decoration used. Generally those labeling systems that rely heavily on paper based substrates such as wet-glue labeling, will compare favourably to laminate structures such as self-adhesive systems that include the cost of the liner and the adhesive. Direct printed containers also tend to have high price per thousand, but offer significant benefits on the filling line because no further labeling operation in required. Printing and conversion costs such as repro, tooling,

PRICE PER '000 LABELS

Basic spec (paper) printed 5 colors

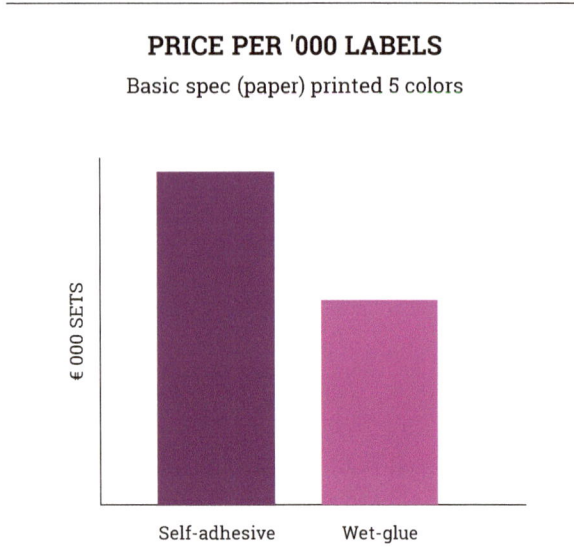

Figure 9.1 Cost per '000 label comparision – self-adhesive versus wet-glue labels

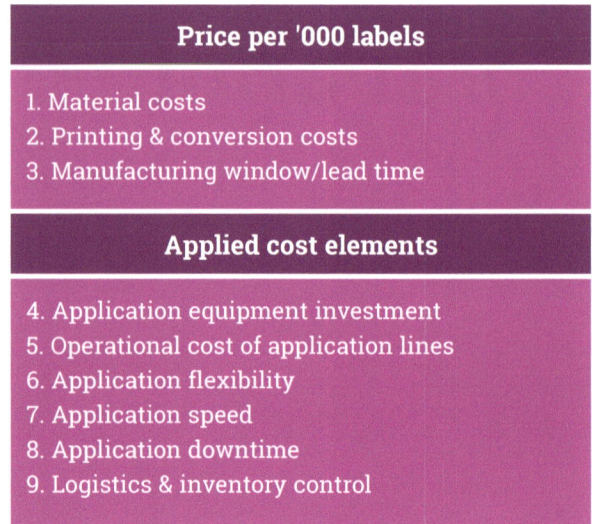

Price per '000 labels

1. Material costs
2. Printing & conversion costs
3. Manufacturing window/lead time

Applied cost elements

4. Application equipment investment
5. Operational cost of application lines
6. Application flexibility
7. Application speed
8. Application downtime
9. Logistics & inventory control

Figure 9.2 Key total applied cost elements

embellishments, finishing and packing materials are also factored into the price per thousand.

Figure 9.1 compares the basic price per thousand (material/print/conversion) for a basic paper label (achievable by wet-glue and self-adhesive methods only). As one might expect when making a comparison on a price per thousand basis, the wet-glue label format clearly demonstrates a significant price advantage over self-adhesive.

TAKING A WIDER VIEW

A costing model that takes a total applied cost approach offers a new perspective on the cost of decorating a pack. Total applied costing includes a host of factors such as the cost and efficiency of label application, investment in capital equipment and machinery change parts, logistics and inventory control.

Details of those factors that can be included in a total cost model are provided in Figure 9.2.

APPLICATION EQUIPMENT

Application equipment is a major cost factor contributing to the cost of labeling a pack.

Investment costs for wet-glue and self-adhesive application equipment tend to be comparable. Sleeving involves both an application system and additional shrink tunnels and although cost effective may require more than one applicator per line to achieve equivalent application speeds.

Consideration must be also be given to the high capital expenditure required for in-mold labeling equipment (by the molder).

When a number of different labels are being applied on the same line then the picture is more complex. Down-time incurred when changing between variants will be reflected in increased labor costs and loss of productivity. Self-adhesive is generally considered to be the most versatile and flexible label system when it comes to multi-variant labeling because reels can be easily changed, down times are less and they often involve fewer people on the line. Digital direct decoration methods offer significant advantages for jobs that have a large number of variants or require personalization.

Rapid changing with in-mold and some direct print decoration can be carried out on the filling line but

A TOTAL APPLIED COST CASE STUDY

Comparative costing model - self-adhesive versus wet-glue labeling

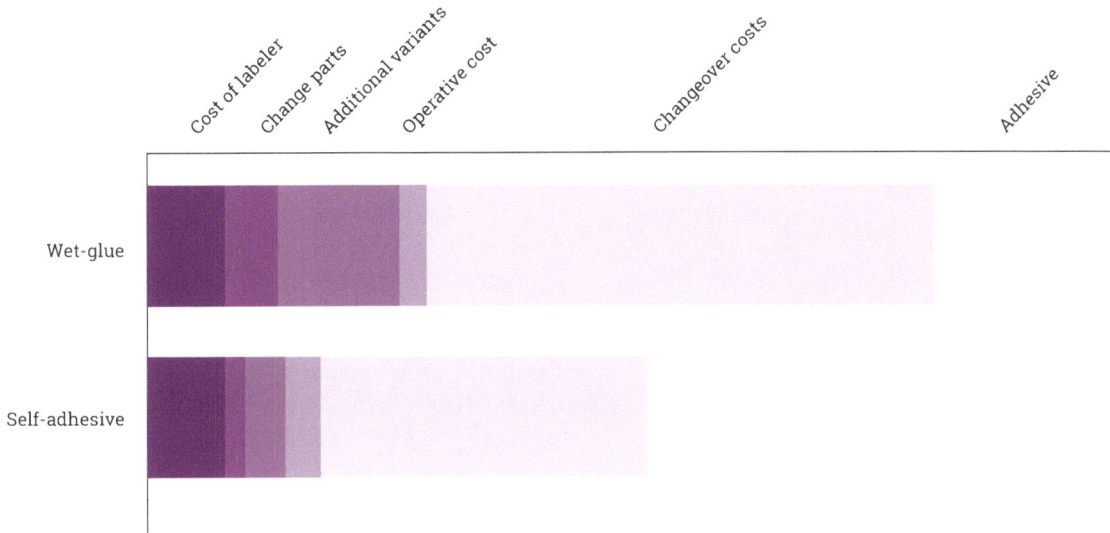

Figure 9.3 Comparative Costing Model – Self-adhesive versus wet-glue labeling

can require considerable logistic and inventory control (resulting from the storage and use of pre-decorated containers).

Adopting flexible packaging formats can offer considerable cost benefits in that no secondary labeling is required, with the primary packaging doubling as the decorative carrier of branding. This decorative system however, is commonly integrated with the product filling operation.

CUSTOMER CASE STUDY

A revealing case study into the cost of labeling also stemmed from the work conducted by 4impression on behalf of FINAT.

The analysis looked at the actual calculations and the decision making process adopted by an end user in the high quality beverages sector.

The decision to select either wet-glue or self-adhesive labeling for their containers was based on an assessment of applicator costs, change parts for each label variant, adhesive costs, operative costs and changeover costs. A breakdown of comparative costs for this user are highlighted in Figure 9.3

Although the cost of the applicator for both systems was similar, the analysis clearly showed that changeover and downtime costs relating to switching between variants was the biggest single contributor to total applied cost and these weighed heavily against the wet-glue format.

In addition, the cost of change parts for label applicators were also a major factor, with the cost of change parts for the wet-glue system being almost twice that of the self-adhesive format. The additional adhesive cost for wet-glue was also taken into account.

On a total applied cost basis self-adhesive was a clear winner in this case.

The practicalities of applying the label were a major contributor to costs. A wet-glue operation often involves longer start-up times in the bottling hall and more change parts for different sizes and shapes of labels, whilst a self-adhesive line has a quicker changeover and can involve fewer people on the line. In this particular case study the user had to accommodate a large number of variant changes. In situations where fewer variants are used the relative costings would change and perhaps a different method of labeling preferred.

DESIGN CONSIDERATIONS

It must be remembered that there are a number of non-cost based factors that can often eliminate certain decoration methods from any effective cost comparison. The marketing or design brief may, for example, dictate that the label contains gold hot foiling which may eliminate shrink sleeving from the equation.

On the other hand a requirement to achieve decoration with 360 degree graphics and with total encapsulation of a pack would strongly favour shrink sleeving and effectively eliminate all other decoration methods.

The higher the graphic content and the wider the range of surface finishes and embellishments required, the more likely it is that self-adhesive or wet-glue labeling will be selected as the optimum decoration method.

PERFORMANCE CONSIDERATIONS

Other practical considerations such as product resistance, chemical/water resistance, pack durability, and even pack safety often pull rank over cost issues.

If a decorative system cannot meet the required specification then it cannot be considered. A shrink sleeve on a glass bottle for instance may offer practical solutions to fragment retention, light weighting and the surface protection of the print (clear sleeves are often reverse printed) and may therefore be the only solution on offer to solve this problem. Alternatively a self-adhesive label may be the only option if the pack has to be sterilized or autoclaved.

Pre-labeled containers such as a direct decorated or in-mold container may be the best way of optimizing speeds on the filling line.

Only when two or more decoration systems are able to meet the specification, can comparative costing take place.

LEAD TIMES

The 'manufacturing window' is an important consideration when choosing the method of decoration because it will impinge on a suppliers' ability to meet lead times. When deadlines are critical, for example in the case of a product launch, marketing and logistical considerations often override cost considerations.

The 'manufacturing window' is extended particularly when the label specification includes some form of embellishment e.g. hot foil-stamping, embossing, lamination.

Sheet-fed, wet-glue label manufacturing requires a separate machine pass for each embellishment process extending the manufacturing window considerably. By comparison roll-fed processes can produce the printing and embellishing in 'one pass' considerably reducing lead times.

Direct digital printing with its print on demand capability has significant advantages where lead-times are tight.

THE ENVIRONMENTAL COST OF LABELING

It is anticipated that cost evaluations of label systems will increasingly have an environmental dimension attached to them.

End users taking a holistic view will want to find ways to reduce the financial burden of packaging waste. They will inevitably favour labeling formats that facilitate the recovery and recycling of packaging waste.

Any converter that can present a total cost model that takes account of environmental impact, will be at a competitive advantage to their counterparts.

Labeling systems that are best able to facilitate total pack recyclability, cut down on application and process waste and help reduce excess packaging will become the solutions selected by end-users. Issues that could come into play are the difficulties and costs involved in removing labels or sleeves in order to re-use or recycle packs and the costs attached to the disposal of backing liners on self-adhesive systems.

The ability of the label system to reduce excess packaging in the supply chain could also be significant. For example a self-adhesive label-leaflet may offer a way for users to eliminate secondary packaging such as cartons. The graphics and information currently carried on an outer carton could now be incorporated within a label-leaflet.

Clearly there is a need to consider the total impact of labeling in the packaging supply chain and the role that the label plays in waste reduction and its influence on the ability of the final pack or product to be recycled.

LOOK AT THE WHOLE PICTURE

Often there is more than one way to decorate a product and each solution must be carefully evaluated before a rational decision can be made. Rather than looking at any one or more criteria in isolation, a more appropriate approach is one that evaluates the total cost of labeling a pack to the required quality, performance and application speed, within a specified time frame.

Some customers will continue to look at the price of the label and not the cost of the label, but it is clear that users are coming under pressure to consider a much wider range of cost factors when selecting the most appropriate method of decorating their product. The environmental cost of labeling is a case in point.

For those who are not currently taking a wider view it is only a matter of time before they have to bend to market drivers. In the meantime it is often in the interest of converters and suppliers to present a lifecycle view of the labeling process so that all costs are reviewed and fair and informed comparisons between labeling and other decoration systems can be made.

Index

www.ingramcontent.com/pod-product-compliance
Lightning Source LLC
Chambersburg PA
CBHW041722210326
41598CB00007B/742